AGROECOLOGY IN PRACTICE

From Local Initiatives to Global Scaling Through Video

Reviewer endorsements for *Agroecology in Practice*

A delightful exposure to practical agroecology principles and practices through charming prose, linked to engaging videos. The book's scope – from key technical topics to the human dimensions, to the power of video to share knowledge and experience – makes this a particularly satisfying volume.
Rebecca Nelson, Professor, Cornell University, MacArthur Fellowship Awardee

The broad and many forms of more sustainable agriculture are rich with creativity and effective methods for producing more food at the same time as benefitting people and nature.
Jules Pretty, Emeritus Professor of Environment &
Society at the University of Essex, author of *Agri-Culture*

An absolutely excellent, diverse store of knowledge, collected from around the world, about how to inexpensively improve farmers' agricultural production while maintaining soil quality and overcoming droughts.
Roland Bunch, author of *Two Ears of Corn*

One of the most comprehensive and valuable books on agroecological systems.
Andre Leu, International Director of Regeneration International,
author of *Growing Life, Poisoning our Children,*
and *The Myths of Safe Pesticides*

An inspiring and informative vision for reforming global agriculture to better serve people and planet.
David R. Montgomery, author of *Dirt: The Erosion of
Civilizations* and *Growing a Revolution*

Agroecology in Practice

From Local Initiatives to Global Scaling Through Video

Jeffery W. Bentley

Paul Van Mele

AGRO insight
communicating agriculture

CABI

CABI is a trading name of CAB International

CABI
Nosworthy Way
Wallingford
Oxfordshire OX10 8DE
UK

CABI
200 Portland Street
Boston
MA 02114
USA

Tel: +44 (0)1491 832111
E-mail: info@cabi.org
Website: www.cabi.org

T: +1 (617)682-9015
E-mail: cabi-nao@cabi.org

The views expressed in this publication are those of the author(s) and do not necessarily represent those of, and should not be attributed to, CAB International (CABI). Any images, figures and tables not otherwise attributed are the author(s)' own. References to internet websites (URLs) were accurate at the time of writing.
CAB International and, where different, the copyright owner shall not be liable for technical or other errors or omissions contained herein. The information is supplied without obligation and on the understanding that any person who acts upon it, or otherwise changes their position in reliance thereon, does so entirely at their own risk. Information supplied is neither intended nor implied to be a substitute for professional advice. The reader/user accepts all risks and responsibility for losses, damages, costs and other consequences resulting directly or indirectly from using this information.

CABI's Terms and Conditions, including its full disclaimer, may be found at https://www.cabidigitallibrary.org/terms-and-conditions.

A catalogue record for this book is available from the British Library, London, UK.

ISBN-13: 9781800628762 (hardback)
9781800628779 (paperback)
9781800628786 (ePDF)
9781800628793 (ePub)

DOI: 10.1079/9781800628793.0000

Commissioning Editor: Ward Cooper
Editorial Assistant: Helen Elliott
Production Editor: Rosie Hayden

Typeset by Exeter Premedia Services Pvt Ltd, Chennai, India
Printed and bound in the USA by Integrated Books International, Dulles, Virginia

Contents

Foreword

We know that our current food systems are not sustainable and that the multiple crises we face as a society threaten the sustainability of our common future and the future of our commons.

The imperative to transform towards sustainable food systems cannot be overstated. Agroecology offers a transformative pathway in this direction for it addresses the entire food system – from production to consumption. It is changing social relations, empowering family farmers and local communities, adding value locally and privileging short value chains that link consumers and producers.

Holistic and integrated, agroecology champions gender equality and social justice, while nurturing identity and culture. This book is a testament to how local communities are innovating to address head-on the constraints they face not only to survive, but also to thrive. Not only are these presented through this book, but they are also captured in videos (which are referenced after each story) through the years of work of both Paul and Jeffery and the dedicated team and partners of Access Agriculture.

Local context matters. These stories show how local innovators are experimenting, adopting and adapting – all the time – to determine what works best given their biophysical and ecological environments as well as their local socio-cultural contexts. Innovation, after all, is what farmers do when scientists and other development agents are not there.

Globally, there is some momentum and interest in agroecology as an effective pathway towards food systems transformation, contributing to addressing converging socio-ecological crises. Paradoxically, funding for further development and implementation of agroecology in various territories or associated research and development interventions is not (yet) commensurate with this recognition.

Facilitating the co-creation and exchange of knowledge and practice in agroecology is crucial to sustain this interest, to gain more political and funding support, and to allow it to spread even further.

Oliver Oliveros
Coordinator of the Agroecology Coalition

Preface

Stephen Gliessman defines agroecology as:

> The integration of research, education, action and change that brings sustainability to all parts of the food system: ecological, economic, and social. It's transdisciplinary in that it values all forms of knowledge and experience in food system change. It's participatory in that it requires the involvement of all stakeholders from the farm to the table and everyone in between. And it is action-oriented because it confronts the economic and political power structures of the current industrial food system with alternative social structures and policy action. The approach is grounded in ecological thinking where a holistic, systems-level understanding of food system sustainability is required. (Gliessman, 2018, p. 12)

Practical agroecology 'values the local, empirical, and indigenous knowledge of farmers, and the sharing of this knowledge' (Gliessman, 2018, p. 12). That is what this book is about, the men and women who regenerate their soil, manage their pests biologically, who save on water, sell food in their communities and care for native crop varieties. For example, some farmers in Bolivia are building up the soil by growing vegetables under a mix of forest trees and apples, selling the organic fruit locally. During COVID-19, a community group in India began to take orders for agroecological produce on social media, and deliver the goods to local, urban households. In many of these experiences the farmers benefit from the contacts, organizational skills and fresh ideas of enlightened outsiders: development professionals, agronomists, researchers, extensionists and entrepreneurs.

Agroecology must build upon yet go beyond these examples, helping many other farmers to learn from pilot experiences. In the jargon, this is called 'scaling up'. Most countries have a cadre of extension agents, who are usually underpaid and stretched thin. For example, the ratio of extensionists to farmers is 1:10,000 in Ghana, 1:4340 in Mexico and 1:9470 in India (Yang and Ou, 2022). Some of these extension agents are still being directed to teach farmers to use agrochemicals, but others do promote agroecology, like the ones with the Mountain Institute in the Cordillera Blanca of Peru, described in this book, who teach farmers how to use rotational grazing to improve the health of soil and livestock. Extensionists alone will never be able to reach all of the smallholder farmers who need more information about agroecology. This book illustrates that an innovative, effective way to share this knowledge with farmers across countries and continents is through videos.

Both authors had a long experience in agroecology before becoming interested in videos. Jeff Bentley learned from smallholders in Honduras about wasps, fire ants and other enemies of maize and bean pests, to encourage biological pest control (Bentley, 1989). For the past four decades, Jeff has seen many inspiring community-driven initiatives in Latin America, South Asia and Africa. As an agricultural anthropologist with a deep interest in farmers' knowledge and local food cultures, Jeff has been able to see how farmers, agronomists and people of other professions interacted in order to make food systems more resilient to climate change and better cope with the destabilizing effects of globalization.

In the early 1990s, trained as a tropical agronomist, Paul Van Mele became intrigued when he saw farmers in Malaysia keep their goats and sheep healthy with herbal medicines. Later on, spending some time in Sri Lanka, he saw how farmers creatively managed highly diverse agroforestry systems. Vietnamese farmers taught Paul about controlling orchard pests with weaver ants, as a functional alternative to insecticides (Van Mele and Cuc, 2007). The many, encouraging experiences Paul saw in developing countries stood in sharp contrast with the way chemical-intensive agriculture was undermining farmers' autonomy back home in Europe. The rise of supermarkets started to put small, local food stores in a stranglehold.

Thinking about how the positive examples from developing countries could be amplified, in 2003, Paul started making videos with video producers from Countrywise Communication (UK) and agricultural scientists at RDA (Rural Development Academy) in Bogra, Bangladesh. With farmers in Mariya village, they made four videos on how to clean, dry and store healthy rice seed. The techniques filmed, such as making simple tables to dry seed, were developed by researchers and smallholders working together.

A few years later, Paul found himself working as the Learning and Innovation specialist at the Africa Rice Center (AfricaRice), an international research organization in Benin. When AfricaRice asked Paul to make some videos for rice farmers, he said that there were already four suitable videos on seed health, filmed in Bangladesh. The videos just needed to be translated into key African languages.

There was pushback from colleagues who thought that the videos needed to be filmed over again in West Africa. Paul and his wife, Marcella Vrolijks, developed several more farmer training videos, but on other topics. Strengthened by the strong requests he received from rural women in various African countries, Paul stood his ground, and over a period of 5 years the rice videos, including the four from Bangladesh, were translated into 40 African languages. The translated videos worked. We later found that Nigerian farmers identified with videos from Bangladesh as easily as with the ones filmed in nearby Mali (Bentley and Van Mele, 2011). Farmers in Uganda who watched the videos instantly recognized the Bangladeshi farmers as authentic smallholders by the 'mud on their legs' (Bentley *et al.*, 2013).

In 2010, Paul started Agro-Insight, a company that makes videos and other communication products for farmers. Two years later, together with his friends and colleagues from Countrywise Communication, he co-founded the non-profit organization Access Agriculture to facilitate global and local sharing of training videos. Access Agriculture created a specialized online platform that allow videos, audio files and factsheets to be downloaded for free. The videos show farmers practising agroecology. These farmers had often honed their ideas by working closely with research and extension. Around the world, other extensionists, educators and farmers themselves download the videos and use the information to spark their own experiments (Bentley *et al.*, 2019, 2022; Van Mele *et al.*, 2010).

Paul Van Mele and Jeff Bentley have worked together for more than 20 years, in countries as diverse as Benin, Bolivia and Bangladesh, influencing each other in many ways, leading up to this book, *Agroecology in Practice*.

Part 1 of this book outlines the disasters provoked by industrial agriculture. Part 2 describes diverse and ecological farming as a viable alternative. Part 3 covers some of the social innovations that are required to build healthy food systems, and Part 4 is about communicating with farmers through video. In each Part, various chapters cover specific themes that are illustrated by succinct, real-life stories that offer insights from Africa, Latin America, Asia, Europe and the USA. These examples from different parts of our planet illustrate the commonalities that connect us all. With each story we provide a link to a related video, featuring real farmers. We hope the reader will share in our concern for farming's present situation, but also in our hope for a better future. We hope that this book will offer some insights for all those concerned with the Earth we live on, and the food we eat, today and tomorrow.

References

Bentley, J.W. (1989) What farmers don't know can't help them: The strengths and weaknesses of indigenous technical knowledge in Honduras. *Agriculture and Human Values* 6(3), 25–31.

Bentley, J.W. and Van Mele, P. (2011) Sharing ideas between cultures with videos. *International Journal of Agricultural Sustainability* 9, 258–263.

Bentley, J.W., Van Mele, P. and Musimami, G. (2013) The Mud on Their Legs – Farmer to Farmer Videos in Uganda. MEAS Case Study # 3. Available at: https://agroinsight.com/downloads/Articles-Agricultural-Extension/2013_AE5_MEAS-CS-Uganda-Farmer-to-Farmer-Videos-Uganda-BentleyJ- and%20PVanMele-July%202013.pdf (accessed 16 October 2024).

Bentley, J.W., Van Mele, P., Barres, N.F., Okry, F. and Wanvoeke, J. (2019) Smallholders download and share videos from the internet to learn about sustainable agriculture. *International Journal of Agricultural Sustainability* 17, 92–107.

Bentley, J.W., Van Mele, P., Chadare, F. and Chander, M. (2022) Videos on agroecology for a global audience of farmers: An online survey of Access Agriculture. *International Journal of Agricultural Sustainability* 20, 1100–1116.

Gliessman, S. (2018) Defining Agroecology. *Agroecology and Sustainable Food Systems* 42, 599–600.

Van Mele, P. and Cuc, N.T.T. (2007) *Ants as Friends: Improving your Tree Crops with Weaver Ants*. CAB International, Wallingford, UK.

Van Mele, P., Wanvoeke, J. and Zossou, E. (2010) Enhancing rural learning, linkages and institutions: The rice videos in Africa. *Development in Practice* 20(3), 414–421.

Yang, P. and Ou, Y. (2022) *Transforming Public Agricultural Extension and Advisory Service Systems in Smallholder Farming: Status Quo, Gaps, Way Forward*. FAO, Rome, Italy.

Acknowledgements

We thank the following partners and colleagues:

Access Agriculture. Josephine Rogers, Phil Malone and all the personnel, for their help and support with video hosting, workshops and follow-up studies such as the ones mentioned in the story 'Videos to Encourage Agroecological Food Systems'.

AGRECOL Andes. Alberto Cárdenas, Alexander Espinoza, María Omonte and Mariana Alem for facilitating fieldwork, for the 'Choosing to Farm' story and others.

Agroecología y Fe. Germán Vargas, Marcelina Alarcón and Freddy Vargas for 'Farming with Trees' and other stories.

Asociación de Organizaciones de los Cuchumatanes (ASOCUCH) for support in Guatemala: for the 'Seed Fairs' story.

Blue Gold Project, financed by the government of the Netherlands to improve water management in Bangladesh, for facilitating work on 'Eating an Old Friend'.

Christian Commission for Development in Bangladesh. Shamiran Biswas, for 'Ashes to Aphids', and Abu Sharif Md. Mahbub-E-Kibria 'Kibria' for help with 'Tomatoes Good Enough to Eat'.

EkoRural. Guadalupe Padilla, Ross Borja and Pedro Oyarzún for their collaboration on 'Enlightened Agroecology', 'What Is a Women's Association About?' and other stories.

Grupo Yanapai. Edgar Olivera, Jhon Huaraca and Raúl Ccanto, for their help with 'Native Potatoes', '200 Guinea Pigs', 'Killing the Soil with Chemicals', and other work.

IRD (*Institut de Recherche pour le Développement*). Diego Mina and Mayra Coro for their help with 'More Insects, Fewer Pests'.

Mountain Institute. Vidal Rondán for facilitating fieldwork on 'Moveable Pasture'.

Plataforma Regional de Suelos in Cochabamba, for hosting 'Friendly Germs'.

PROINPA (*Fundación para la Promoción e Investigación de Productos Andinos*). Alejandro Bonifacio, Eliseo Mamani, Jorge Blajos, Genaro Aroni, Milton Villca, Juan Almanza, Reinaldo Quispe, Rhimer Gonzales and Rolo Oros for facilitating fieldwork with 'Wind Erosion and the Great Quinoa Disaster', 'Don't Eat the Peals', 'Awakening the Seeds', 'United Women of Morochata', 'Seeing the Life in the Soil' and other stories.

PROSUCO (*Promoción de la Sustentabilidad y Conocimientos Compartidos*). Eliodoro Ballivián, María Quispe, Maya Apaza, Renato Pardo, Roly Cota and Sonia Laura, for their support with 'Harsh and Healthy', 'Zoom to Titicaca', 'Organic Leaf Fertilizer' and other stories.

SWISSAID. Fernando Jácome and Oscar Quillupangui for 'When Local Authorities Support Agroecology'.

We also thank all of the people mentioned by name in the text of this book for their generosity with their time and information.

Thanks to Eric Boa for his encouragement and for his valuable comments on many parts of this book.

We gratefully acknowledge the following financial support:

The videos made in Peru, Ecuador and Bolivia (and associated fieldwork) were made possible with the kind support of Global Collaboration for Resilient Food Systems (CRFS) of the McKnight Foundation.

For many years, the Swiss Agency for Development and Cooperation (SDC) has offered crucial support to Access Agriculture to build local capacities to produce quality farmer-to-farmer training videos and translate these into many local languages.

Many other organizations and donors supported the development of the quality learning videos referred to in this book.

Introduction

Bolivia ran out of diesel fuel recently. The whole country was suddenly out. Trucks began to line up at service stations, until there were none left on the road. The drivers courteously left gaps in the queue, to avoid blocking side streets and driveways. But the lines eventually reached for many kilometres, often on each side of the highway. Many of the trucks were carrying food: soy for export, groceries from neighbouring countries and a lot of internally traded fresh produce. It was all tied up in knots. Panicked shoppers at one government-run supermarket were lined up from the cash register all the way to the back of the store.

Thankfully, a few days later the trucks were being refuelled and the crisis was over, at least temporarily. But it was a frightening reminder that even lower-middle-income countries are tangled up in a precarious food network that is hooked on chemicals, machinery and lots of fossil fuel. Wealthy countries may appear to be too well organized to have such a mess, but of course, food webs there reach even further, and are even more complex than they are in the Global South. The whole network is vulnerable.

A visit to a supermarket in a country like the United States gives shoppers an image of never-ending abundance. But it's based on farming systems that are rapidly eroding the soil, drawing down the groundwater and throwing family farmers off the land (Philpott, 2020). Recent advances in computer and digital technology give the impression of rapid, almost magical innovation. This masks the fact that other sectors have changed much less since the 1970s. Agriculture and transportation demand ever greater amounts of non-renewable energy, fuelling climate change and exposing the food system to greater dependence on outside inputs (Smil, 2023).

This raises the question, can we continue to feed the world without destroying our planet?

Many contemporary food systems are now growing, transporting, selling and consuming food in ways that:

destabilize the climate, degrade our land and water resources, and threaten
critical ecosystems, biodiversity, and people's livelihoods. Agriculture and
food, the world's biggest sector of employment, is predominantly character-
ized by poorly paid, insecure jobs and low prices to farmers. (CGIAR, 2022)

This is according to the CGIAR (formerly the Consultative Group on
International Agricultural Research), a global research partnership for a
food-secure future. The CGIAR Research Strategy mentions agroecology as a
solution to these problems (CGIAR, 2024). Agroecology does offer a promising path forward, integrating ecological, social and economic principles to
create resilient and productive food systems (Gliessman *et al.*, 1998). While
industrial agriculture has dominated global food production, its reliance on
chemical inputs, its contribution to soil erosion and climate change, and its
damage to rural livelihoods have become increasingly unjustifiable. Farmers,
communities and policy makers are seeking alternatives that prioritize ecological balance, social justice and economic viability.

This book explores the transformative power of agroecology through examples of practical, farmer-centred approaches. We visit farmers around the
world who are successfully implementing agroecology, sharing their knowledge and building resilient communities. A cornerstone of agroecology is
the recognition of farmers as both stewards of the land and innovators. The
Campesino a Campesino (farmer-to-farmer) movement, with its emphasis on
peer-to-peer learning, exemplifies this approach (Holt-Giménez, 2006).

Campesino a Campesino builds on the idea that farmer 'promoters'
can best share new ideas, by trying them out on their own farms, adapting them and sharing them with their neighbours, as described by Roland
Bunch (1982) in *Two Ears of Corn*. Innovative farmers share their knowledge
with their peers, fostering a sense of ownership and agency within the community. The educators are fellow farmers, making the new information more
credible than ideas coming from outsiders. Our book builds upon this tradition by showing how new digital technologies allow farmers to share ideas
from their own farm with a global audience of peers.

Agroecology is not merely about reverting to old practices but involves
applying modern scientific principles alongside local knowledge to create
innovative, productive and sustainable agricultural systems (Altieri, 2018).
Our book also stresses farmer innovation that is based on new, scientific
ideas, blended with time-honoured knowledge from the farm community.

Agroecology is also linked to food sovereignty: the right of peoples
to healthy and culturally appropriate food produced through ecologically
sound and sustainable methods. Agroecology aims to conserve biodiversity:
the variety of life on Earth, that is, diverse genes, species and ecosystems
(Perfecto *et al.*, 2019). The case studies in this book show that smallholders
need to be the guardians of their own agrobiodiversity. Farmers' rights to
grow their own crop varieties must be asserted. Rights over seed and varieties are the most basic kind of food sovereignty, and crucial for managing the
genetic diversity needed to feed the world into the future.

Agroecology recognizes the social life of agriculture, especially the collaboration among stakeholders (Tittonell, 2023). The transition to agroecology will not happen only on individual plots or farms, but must also include support from policy and research at communal and larger scales. Farmers need consumers to make a living, and consumers rely on farmers for healthy food. We emphasize this symbiotic relationship in our book, with cases on farmer associations, and on the need to form closer ties between farmers and consumers. We discuss short food chains as a counter-weight to large corporations in the food system.

By showcasing successful examples of agroecology and by providing practical guidance, we hope to inspire and empower farmers and other food producers, researchers, entrepreneurs, policy makers and consumers to seek new ways to shape our food systems, while protecting the world we live in.

References

Altieri, M.A. (2018) *Agroecology: The Science of Sustainable Agriculture*. CRC Press, Boca Raton, Florida.

Bunch, R. (1982) *Two Ears of Corn: A Guide to People-Centered Agricultural Improvement*. World Neighbors, Oklahoma City, Oklahoma.

CGIAR (2022) CGIAR Research and Innovation Strategy: Transforming Food, Land and Water Systems in a Climate Crisis. CGIAR System Organization, Montpellier, France.

CGIAR (2024) CGIAR website. Available at: www.cgiar.org (accessed 16 October 2024).

Gliessman, S.R., Engles, E. and Krieger, R. (1998) *Agroecology: Ecological Processes in Sustainable Agriculture*. CRC Press, Boca Raton, Florida.

Holt-Giménez, E. (2006) *Campesino a Campesino: Voices from Latin America's Farmer to Farmer Movement for Sustainable Agriculture*. Food First Books, Oakland, California.

Perfecto, I., Vandermeer, J. and Wright, A. (2019) *Nature's Matrix: Linking Agriculture, Biodiversity Conservation and Food Sovereignty* (2nd edn). Routledge, Abingdon, UK.

Philpott, T. (2020) *Perilous Bounty: The Looming Collapse of American Farming and How We Can Prevent It*. Bloomsbury Publishing, New York.

Smil, H. (2023) *Invention and Innovation: A Brief History of Hype and Failure*. The MIT Press, Cambridge, Massachusetts.

Tittonell, P. (2023) *A Systems Approach to Agroecology*. Springer Nature, Berlin, Germany.

PART 1

Problems with Chemical-Intensive, Industrial Agriculture

We are cautiously optimistic about agriculture and the environment. Farming is facing great problems, including the destruction of the soil, climate change and the abuse of agrochemicals, antibiotics and veterinary drugs (McKenna, 2017; Philpott, 2020). Supermarkets and global fast-food chains are undermining local food cultures, which depend on local farming and gardening. Farmers can be driven out of business by thoughtless policies as farm sizes steadily increase in the developed world (Lowder *et al.*, 2016; Wetherington, 2021). Agroecology respectfully builds on local knowledge and farmer experiments (Pimbert, 2018, p. 12). Although most of this book is light on gloom and doom, Part 1 starts by recognizing some of the serious challenges that farming faces, before moving on to solutions in Parts 2–4.

References

Lowder, S.K., Skoet, J. and Raney, T. (2016) The number, size, and distribution of farms, smallholder farms, and family farms worldwide. *World Development* 87, 16–29.

McKenna, M. (2017) *Big Chicken: The Incredible Story of How Antibiotics Created Modern Agriculture and Changed the Way the World Eats*. National Geographic, Washington, DC.

Philpott, T. (2020) *Perilous Bounty: The Looming Collapse of American Farming and How We Can Prevent It*. Bloomsbury Publishing, New York.

Pimbert, M.P. (2018) *Food Sovereignty, Agroecology and Biocultural Diversity: Constructing and Contesting Knowledge*. Routledge, Abingdon, UK.

Wetherington, M.V. (2021) *American Agriculture: From Farm Families to Agribusiness*. Rowman and Littlefield, Lanham, Maryland.

1

Environmental Degradation, Climate Change and Health Risks

Abstract

Global food systems grapple with mounting pressures including soil erosion, biodiversity loss, water scarcity and climate change. Conventional agriculture's heavy reliance on chemicals exacerbates these challenges, leading to environmental degradation, public health concerns and economic instability for farmers. Case studies reveal the far-reaching consequences of unsustainable practices, such as the quinoa boom's destruction of ecosystems and the widespread overuse of herbicides. Agroecology offers a transformative approach by prioritizing ecosystem health, social equity and climate resilience. By adopting agroecological practices, we can build more sustainable and resilient food systems.

Wind Erosion and the Great Quinoa Disaster

Bolivian agronomist Genaro Aroni first told me how quinoa was destroying the south-western Bolivian landscape in 2010, when he came to Cochabamba for a writing class I was teaching. I could hardly believe his stories of how the soil was being completely blown away, but he was right.

In 2018 I got my chance to see this Bolivian Dust Bowl, when Paul Van Mele, Marcella Vrolijks and I were making videos with Agro-Insight for the McKnight Foundation. Together with Milton Villca, an agronomist from the Foundation for the Promotion and Research of Andean Products (PROINPA – *Fundación para la Promoción e Investigación de Productos Andinos*), we met Genaro in Uyuni, near the famous salt flats of Bolivia. Genaro told us that he had worked with quinoa for 41 years, and had witnessed the dramatic change from mundane local staple to global health food. He began explaining what had happened.

© Jeffery W. Bentley and Paul Van Mele 2025. *Agroecology in Practice: From Local Initiatives to Global Scaling Through Video* (J.W. Bentley and P. Van Mele) DOI: 10.1079/9781800628793.0001

When Genaro was a kid, growing up in the 1950s, the whole area around Uyuni, in the arid southern Altiplano, was covered in natural vegetation. People grew small plots of quinoa on the low hills, among native shrubs and other plants. Quinoa was just about the only crop that would survive the dry climate at 3600 m above sea level. The llamas roamed the flat lands surrounding the hills, growing fat on the native brush. In April, the owners would pack the llamas with salt blocks cut from the Uyuni Salt Flats (the largest dry salt bed in the world) and take the herds to Cochabamba and other lower valleys, to barter salt for maize and other foods that can't be grown on the high plains. The llama herders would trade salt and quinoa for maize, potatoes and *chuño* (dried potatoes) from other farmers, supplementing their diet of dried llama meat and quinoa grain.

Then in the early 1970s, a Belgian project near Uyuni introduced tractors to farmers and began experimenting with quinoa planted in the sandy plains. About the same time, a large-scale farmer further north in Salinas also bought a tractor and began clearing scrublands to plant quinoa.

More and more people started to grow quinoa. The crop thrived on the sandy plains, but as the native brushy vegetation grew scarce, so the numbers of llamas began to decline.

Throughout the early 2000s the price of quinoa increased steadily. When it reached 2500 Bolivianos for 100 lb (US$8 per kilogram) in 2013, many people who had land rights in this high rangeland (the children and grandchildren of elderly farmers) migrated back – or commuted – to the Uyuni area to grow quinoa. Genaro told us that each person would plough as much as 10 ha of native vegetation land to plant the suddenly valuable quinoa crop (Bonifacio *et al.*, 2022).

But by 2014 the quinoa price slipped and by 2015 it crashed to about 350 Bolivianos per hundredweight (US$1 per kilogram), as farmers in the USA and elsewhere began to grow quinoa themselves.

Many Bolivians gave up quinoa farming and went back to the cities. By then the land was so degraded it was difficult to see how it could recover. Still, Genaro is optimistic. He believes that quinoa can be grown sustainably if people grow less of it and use cover crops and crop rotation. Later research by Genaro and colleagues suggested that wild lupins could be rotated with quinoa. Not much else besides quinoa can be farmed at this altitude, with only 150 mm (6 in) of rain per year (Bonifacio *et al.*, 2022).

Milton Villca took us to see some of the devastated farmland around Uyuni. It was worse than I ever imagined. On some abandoned fields, native vegetation was slowly coming back, but many of the plots that had been planted in quinoa looked like a moonscape, or like a white sand beach, minus the ocean (Fig. 1.1).

Farmers would plough and furrow the land with tractors, only to have the fierce winds blow sand over the emerging quinoa plants, smothering them to death. Milton showed one of the few remaining stands of native vegetation. Not coincidentally, this was near the hamlet of Lequepata where some people still herd llamas. Llama herding is still the best way of using this land without destroying it.

Milton showed us how to gather wild seed of the *sup'u t'ula* plant (*Parastrephia lepidophylla*); each bush releases clouds of dust-like seeds,

Fig. 1.1. A one-off quinoa crop wipes out an ancient, dwarf climax forest. Land once covered with healthy grass and brush, reduced to bare sand.

scattered and planted by the wind (Fig. 1.2). Milton and Genaro are teaching villagers to collect these seeds and plant them, to establish windbreaks around their fields, to conserve the soil.

While quinoa has become a celebrated food worldwide, this case is a sobering reminder that rapid commercialization can have unintended, and destructive, consequences for a region suddenly faced with extreme demand for a local food.*

📹 **Related video**

While writing the above section, Agro-Insight and PROINPA filmed the following farmer-learning video, about how to protect the soil from erosion by wind.

Agro-Insight and PROINPA (2018) *Living Windbreaks to Protect the Soil*. Hosted by Access Agriculture. 15 minutes.
www.accessagriculture.org/living-windbreaks-protect-soil

A version of the section above was previously published on www.agroinsight.com in 2018, by Jeff Bentley.

Fig. 1.2. Community members harvesting wild seeds of native plants as a strategy to replant living hedges and reduce wind erosion.

Out of Space

Celebrating the 50th anniversary of the moon landing, a series of weekly broadcasts on Belgian national TV in 2019 nicely illustrated the spirit of the time. One interview with a man on a New York City street drew my particular attention. The interview showed why so many people supported the NASA programme:

> We have screwed up our planet, so if we could find another planet where we can live, we can avoid making the same mistakes.

History has shown repeatedly how the urge to colonize other places has been a response to the declining productivity of the local resource base. In his eye-opening book *Dirt: The Erosion of Civilizations* (2007), Professor David Montgomery from the University of Washington clearly explains the global and local dynamics of land use from a social and historical perspective.

Of the many examples given in his book, I will focus on the most recent one: the growth of industrial agriculture, as the rate of soil erosion has taken on such a dramatic proportion that it would be a crime against humanity not to invest all of our efforts to curb the trend and ensure food production for the next generations.

The First World War triggered various changes affecting agriculture. First, the area of land cultivated on the American Great Plains doubled during the war. The increased wheat production made more exports to Europe possible. Already aware of the risks of soil erosion, in 1933 the US government established an elaborate scheme of farm subsidies to support soil conservation and crop diversification, stabilize farm incomes and provide flexible farm credit. Most farmers took loans to buy expensive machinery. Within a decade, farm debt more than doubled while farm income only rose by a third (Montgomery, 2007).

After the Second World War, military assembly lines were converted for civilian use, paving the way for a tenfold increase in the use of tractors (see also Harper, 2001). By the 1950s, several million tractors were ploughing American fields. On the fragile prairie ecosystem of the Great Plains, soil erosion rapidly took its toll and small farmers were especially hit by the drought in the 1950s. Many farmers were unable to pay back their loans, went bankrupt and moved to cities. The few large farmers who were left increased their farm acreage and grew cash crops to pay off the debt of their labour-saving machinery. By the time Neil Armstrong had put his foot on the moon, four out of ten American farms had disappeared in favour of large, corporate factory farms (Montgomery, 2007).

Just when the end of the Second World War triggered large-scale mechanization, the use of chemical fertilizer also sharply increased. Ammonia factories used to produce ammunition were converted to produce cheap nitrogen fertilizer. Initial yield increases during the Green Revolution stalled and started to decline within two decades. By now, the sobering figures indicate that despite the high-yielding varieties and abundant chemical inputs, productivity in 39% of the area growing maize, rice, wheat and soy bean has stagnated or collapsed. Reliance on purchased annual inputs has increased production costs, which has increased the debt burden of many farmers, and led to farm business failures. Agriculture consumes 30% of our oil use. With the rising oil and natural gas prices it may soon become too expensive to use these dwindling resources to produce fertilizer (Montgomery, 2007).

Armed with fertilizers, farmers thought that manure was no longer needed to fertilize the land. A decline in organic matter in soils further aggravated the vulnerability of soils to erosion. As people saw the soil as a warehouse full of chemical elements that could be replenished ad libitum to feed crops, they ignored the microorganisms that provided a living bridge between organic matter, soil minerals and plants. Unlike plants, most microorganisms lack chlorophyll to conduct photosynthesis, so they require organic matter to feed on (Montgomery, 2007).

A 1995 review reported that each year 12 million hectares of arable land are lost due to soil erosion and land degradation. This is 1% of the available arable soil, per year. The only three regions in the world with deep loess soil for agriculture are the American Midwest, northern Europe and northern China. Today, about a third of China's total cultivated area is seriously eroded by wind and water (Pimentel *et al.*, 1995).

While the plough has been the universal symbol of agriculture for centuries, people have begun to understand the plough's devastating role in soil erosion. By the early 2000s, 60% of farmland in Canada and the USA was already managed with conservation tillage (leaving at least 30% of the field covered with crop residues) or no-till methods. In most other parts of the world, including Europe, ploughing is still common practice and living hedges as windbreaks against erosion are still too often seen as a hindrance for large-scale field operations.

In temperate climates, ploughing gradually depletes the soil of organic matter and it may take a century to lose 10 cm of top soil. This slow rate of degradation is a curse in disguise, as people may not fully grasp the urgency required to take action. However, in tropical countries the already thinner top soil can be depleted of organic matter and lost to erosion in less than a decade. The introduction of tractor-hiring services in West Africa may pose a much higher risk to medium-term food security than climate change, as farmers plough their fields irrespective of the steepness, soil type or cropping system. In Nigeria, soil erosion on cassava-planted hillslopes removes more than 2 cm of top soil per year (Montgomery, 2007).

Despite the overwhelming evidence of the devastating effects of conventional agriculture, the bulk of public research and international development aid is still geared around a model that supports export-oriented agriculture that mines the soils, and chemical-based intensification of food production that benefits large corporations. Farm subsidies and other public investments in support of a more agroecological approach to farming are still sadly insufficient, yet a 2019 report from The High Level Panel of Experts on Food Security and Nutrition concludes that the short-term costs of creating a level playing field for implementing the principles suggested by agroecology may seem high, but the cost of inaction is likely to be much higher (HLPE, 2019).

Inventors, private companies and policy makers are now working on ways to remove oil and natural gas from transportation and agriculture, to mitigate climate change. And whether we are ready for petroleum-free agriculture is a societal decision. When fuel subsidies are phased out in the coming years to curb greenhouse gas emissions, synthetic nitrogen fertilizers are likely to become a technology of the past. With all attention being drawn to curbing the effects of climate change, governments, development agencies and companies across the world also have an urgent responsibility to invest in promoting a more judicious use of the most basic agricultural resource in agriculture: land. We are running out of space and colonizing other planets is not an option for feeding our planet (IPES-Food, 2016).*

▶️ Related video

Farmers in South-east Asia show how to conserve soil, even when growing a root crop, cassava, on sloping land.

Agro-Insight (2016) *Growing Cassava on Sloping Land*. Hosted by Access Agriculture. 12 minutes.
www.accessagriculture.org/growing-cassava-sloping-land

A version of the section above was previously published on www.agroinsight.com in 2019, by Paul Van Mele.

Damaging the Soil and Our Health with Chemical Reductionism

For 150 years, much of the public has become alienated from our food, often not knowing how it was produced, or where. Single-nutrient research papers (Vitamin C cures the common cold! Omega-3 fatty acids reduce the risk of cardiovascular disease!) have eroded our perception of food and provided the basis for food companies to get us to eat more highly processed foods touted as healthier than real food. The work of a few reductionist chemists has had an outsized influence on industrial food production, with devastating effects on soil health and human health.

In 1840, the German scientist Justus von Liebig observed that nitrogen (N), phosphorous (P) and potassium (K) were responsible for crop growth. Later in life, Liebig realized that these macronutrients were far from adequate. He even argued vehemently against the use of nitrogen-based fertilizers for many years, but his progressive insights were largely ignored by the fertilizer industry, which quickly understood that more money can be made by keeping things simple. Occasionally, some micronutrients such as zinc (Zn), magnesium (Mg) or sulfur (S) have been added to blends of fertilizer, but the over-reliance on these chemicals has had a devastating effect on soil ecology, and air and water pollution (Pollan, 2008).

Healthy soils are complicated systems, with a host of macro- and micro-organisms, from earthworms to beneficial fungi and bacteria, interacting with each other to create a living soil. Many universities have shied away from this complex ecology, creating departments of soil physics and soil chemistry, but not ones for soil biology or ecology. Marketing people also favour simplicity. Telling farmers how to apply 120 kg of NPK to grow a crop is easier than educating them on soil ecosystems with all their complex interactions. And these simple recommendations sell more fertilizer.

The nascent food industry was also quick to latch onto simplistic, chemical reductionism. The same Liebig, who promoted nitrogen as plant food, proposed that animal protein (which contains nitrogen) was the fertilizer that makes humans grow (Pollan, 2008).

By 1847 Liebig had invented a beef-based extract, and he went into business with an entrepreneur who bought cheap land on the pampas of Uruguay. From the new port town of Fray Bentos, about 100 miles (160 km) up the Uruguay River from Buenos Aires, Liebig's extract, as thick as molasses, was shipped across the world.

Liebig claimed that his extract contained fats and proteins and could cure typhus and all sorts of digestive disorders. He enlisted physicians and apothecaries to sell his goo. As criticism mounted that there was little nutritional value in his concoctions, the Liebig company changed tack, marketing the product not as a medicine, but as a delicious palliative that could ease a troubled stomach and mind. This change in marketing proved shrewd.

By the early 1870s the extract was a staple in middle-class pantries across Europe. Lest you think we are too smart to be fooled by such chicanery today, the original gooey extract is still sold by the Liebig Benelux company, and meat tea lives on as the bouillon cube. The next time you open a flavour packet that comes with a brick of ramen noodles, you have Liebig to thank (Cansler, 2013).

Liebig and other chemists were influential in reducing food – and the focus of the agri-food industry – to a few, large, simple ingredients. But food is more than a mere combination of nutrients that can be easily measured and prescribed.

While the meat industry has continued to grow, in the early 20th century, dietitians like John Harvey Kellogg strongly opposed eating meat, claiming that animal protein had a devastating effect on the colon (Pollan, 2008). As he laid the foundation for the breakfast cereal industry, Kellogg marketed his products in terms of simple food ingredients: carbohydrates and fibres. While the first packaged breakfast cereals were all whole grain, over the years they have evolved numerous additions, such as dried fruits and lots of refined sugar, and most are now made with white flour. However, they are still marketed as part of a nutritious breakfast.

In his book *In Defense of Food*, Michael Pollan provides ample examples of how over the past 150 years consumers have been made to believe that food can be reduced to calories and simple nutrients. As highly processed foods are filling the shopping baskets of billions of people across the globe, cancers, diabetes and vascular diseases become ever more common (Pollan, 2008).

But the food industry is powerful.

Although soy bean recipes like tofu have been part of a balanced diet for centuries in Asia and whole maize can be made into healthy food like tortillas, both crops are now being subjected to a new reductionism, as they are refined into fat and carbohydrates: 75% of the vegetal oil we use is from soy beans, while more than half of the sweeteners added to our processed food and drinks is high-fructose corn syrup, from maize. Crops that could be part of a healthy diet for people are now either fed to animals in factory farms, or turned into fats and sugar, contributing to the obesity epidemic (Pollan, 2008).

Since the 1970s, the increased focus on maize and soy beans, with their patented varieties, has served three strongly interwoven industries of seed, fertilizer and food manufacturing. Just four companies now dominate seed and agrochemicals globally (Bayer-Monsanto, DowDuPont/Corteva, ChemChina-Syngenta and BASF). While large corporations reap immediate profits, we the tax payers are left to solve the problems they cause in the form of soil erosion, air and water pollution, plummeting biological and food diversity and public health risks.

Fortunately, consumers across the globe are starting to awaken to the risks posed by industrial food production and eating chemically processed food with refined and even artificial ingredients.

The over-reliance on chemicals in agriculture and in food processing is more than an analogy. Both are part of an effort to simplify food systems to a few constituent parts, dominated by a few large players. It has taken the world nearly two centuries to get into this trap, and it will take an effort to get out of it. Agroecology, with its focus on short food supply chains, is pointing the way forward for food that is healthy for the body, mind and society at large.

In March 2021, the European Commission approved an action plan stipulating that 30% of the public funds for agricultural research and innovation have to be in support of organic agriculture. The backlog is huge, so it is timely to see that research will cover, among other things, changing farmers' and consumers' attitudes and behaviours (European Commission, 2021).*

◀ Related video

There are many traditional ways of making nutritious food from soy beans, as we learn in this video from Mali.

AMEDD (Association Malienne d'Éveil au Développement Durable) (2016) *Making a Condiment from Soya Beans*. Hosted by Access Agriculture. 10 minutes. www.accessagriculture.org/making-condiment-soya-beans

---

A version of the section above was previously published on www.agroinsight.com in 2021, by Paul Van Mele.

A Revolution for Our Soil

Degraded soil can be repaired and replenished with nutrients until it produces abundant harvests at lower costs, while removing carbon from the atmosphere and putting it back into the ground. This is the optimistic message of David Montgomery's book *Growing a Revolution* Montgomery (2017).

In many parts of the world, soils have been degraded by frequent ploughing. Ploughing releases a burst of nutrients for the crops and kills weeds, but these benefits are overshadowed by the oxidation of soil carbon, which is released as carbon dioxide into the atmosphere, and exposing the soil to erosion by wind and water (see the above section 'Out of Space' on Montgomery's 2007 book *Dirt: The Erosion of Civilizations*). In the Midwestern USA perhaps half of the original prairie soil, and most of its organic matter, have been lost in little more than a century of conventional tillage. Chemical fertilizers provide the major nutrients (phosphorous, potassium and nitrogen) in the short run, but they undermine the soil's long-term health by suppressing mycorrhizal fungi, which live in symbiosis with plants.

These mycorrhizal fungi feed plants while making glomalin, a protein that binds soil particles together and which is a significant global store of fixed organic carbon and nitrogen. Ploughing destroys the soil structure created by beneficial fungi and their glomalin (Montgomery, 2017).

Montgomery, a professional geologist, explains that most soils don't need chemical fertilizer. They have enough phosphorous, potassium and all the minor nutrients like iron and zinc that plants need, but these minerals are locked up in stone particles and other forms not accessible to the plants. The key to using these nutrients is beneficial microbes, like the mycorrhizal fungi that extract mineral nutrients from rock fragments and help to break down organic matter so plants can use it. Mycorrhizal fungi and soil-dwelling bacteria extract phosphorous and other mineral nutrients from soil particles and rock fragments, and help break down organic matter to soluble nutrients that plants can absorb through their roots. The plants exude proteins and sugars which soil microbes feed on (Montgomery, 2017, p. 46). Predatory arthropods, nematodes and protozoa then feast on the microbes and release the nutrients back to the soil. A diverse soil life makes soil more fertile. Synthetic fertilizers interrupt these interactions, and the mycorrhizal fungi die, so the crop becomes chemical-dependent. Soil that is rich in organic matter (that is, in carbon) is healthier and supports a thriving community of beneficial microorganisms.

But with proper care, soil can be brought back to good health in just a few years. The right techniques can boost soil carbon from 1% (typical of degraded soils) to 4% (as in undisturbed forest) or even up to 6%. There are many such techniques and they go by various names, including 'conservation agriculture', 'agroecology' or 'regenerative agriculture', and they have a few simple principles in common: (i) use cover crops (or mulch) to keep the soil covered all the time; (ii) complex crop rotations of grasses, legumes and other crops; and (iii) no-till, planting seeds directly into the unploughed earth.

Montgomery takes his readers to meet farmers from Kansas to Pennsylvania, from Ghana to Costa Rica, who are practising and profiting from these three principles. Some are organic farmers; others apply small amounts of nitrogen fertilizer directly into the soil, near the seed, where the plant can efficiently take it up. We learn that some use earthworms, while others, like Felicia Echeverría in Costa Rica, make their own brews of beneficial microorganisms, to add life to dead soil. Gabe Brown in North Dakota rotates cattle in small paddocks, on large fields. As the cows graze, they fertilize the soil with manure (Brown, 2018).

Montgomery and soil scientist Rattan Lal estimate that conservation agriculture could offset a third to two thirds of current carbon emissions, by putting organic matter back into the soil, while tilling less and so lowering fuel expenses. Stumbling blocks to adoption of conservation agriculture include subsidies and crop insurance that keep farmers ploughing and dependent on chemical fertilizer. Another is formal research, which continues to privilege studies of products that companies can sell: chemical solutions to biological problems, as Montgomery puts it. Only 2% of US agricultural research goes to regenerative agriculture (and only 1% globally). Much of the innovation to revive the soil is driven not by funded research but by the farmers themselves, who have shown that conservation agriculture, agroecology, regenerative agriculture and permaculture can be more productive, with fewer pest problems. Conservation agriculture saves on expenses for inputs, so it is more profitable than conventional tillage agriculture, an idea echoed by Brown (2018). Properly conserved soil has little erosion; it soaks up the rain in wet years and holds the moisture for drought years.

Montgomery (2017) is concerned that when large-scale, industrialized farmers convert from tillage to conservation agriculture there must be a transition period when profits sag, before the soil improves enough to bring yield back up. He fears that this can discourage farmers from switching to conservation agriculture. Yet I am sure that the farmers themselves will work this out. As the natural experimenters that they are, farmers can try ecological farming practices with reduced tillage, first on one field, or on part of one, gradually creating the practices they need, one plot at a time (Hansson, 2019). The good news is that regenerative agriculture can be adopted on large farms or small ones, conventional or organic, mechanized or not. Farming can rebuild the soil, and does not need to destroy it.*

Related video

This video is the first of a 12-part series on sustainable land management.

Countrywise Communication and CIS Vrije Universiteit Amsterdam (2016) *Sustainable Land Management: Introduction.* Hosted by Access Agriculture. 7 minutes. www.accessagriculture.org/slm00-introduction

---

A version of the section above was previously published on www.agroinsight.com in 2020, by Jeff Bentley.

Roundup®: Ready to Move On?

At our local garden shop in north-eastern Belgium, I recently overheard a conversation between the shopkeeper and a young customer, who asked about the herbicide Roundup®. Since glyphosate, the active ingredient in Roundup®, was banned in Belgium for home use (Phys.Org, 2023), a new glyphosate-free Roundup® is now aggressively promoted in garden centres. The original Roundup® can only be used for professional farming, so the shopkeeper explained that her husband is often asked to go and spray people's ornamental home gardens. Even chemical habits can be hard to kick.

When it is my turn at the counter (I am looking for organic chicken feed), I tell the shopkeeper that I just returned from an international conference where American professors revealed how various ingredients of Roundup® can be related to male infertility, cancer, Alzheimer's disease and at least 40 other human diseases. She took in the information without being shocked

Fig. 1.3. Farmers often do not read or ignore safety instructions when using pesticides.

and countered that many people have recently resorted to homemade remedies like vinegar to kill weeds, which she preposterously claimed did much more harm to the soil than commercial products. Some people who sell chemicals, even at the retail level, can become jaded about their dangers, and prone to spreading disinformation.

In our long experience, both in developed and developing countries, very few people think it is necessary to protect themselves when spraying pesticides. People either cannot read, fail to make the effort to read the label or ignore the risks (Figs 1.3 and 1.4).

While debates on cause-effect relationships can last for decades (the tobacco lobby successfully denied the carcinogenic effects of tobacco for decades, knowing all the while that smoking was a killer), the scientific presentations at the international conference I attended also revealed the shortcomings of official systems that have been put in place to protect our public health. For one, toxicity trials before new products are released only look at short-time effects, whereas diseases of mice (and humans) often show symptoms after years of chronic exposure, as the toxins build up in the body. Equally important, official tests are only done on the active ingredient, not on the full product as it is sold and used (Defarge *et al.*, 2018).

Protected by intellectual property rights, companies are not obliged to reveal and list the ingredients of the inert material that makes up the bulk of

Fig. 1.4. In many countries, children are exposed to toxic pesticides.

herbicides and pesticides. Laboratory tests showed that one of the ingredients in Roundup® is arsenic, which is at least 1000 times more toxic than the glyphosate itself. In short, the glyphosate-free Roundup® is still as toxic as before, only this is not shown in official tests.

Environmental damage, including pollution, soil erosion and biodiversity loss, is hard to measure in simple economic terms. As we mention in the section 'What Counts in Agroecology' (see Chapter 14, this volume), environmental costs are often seen as 'externalities' and are ignored when calculating the cost–benefit of farms. This has given conventional farming an unfair advantage over organic or agroecological farming (HLPE, 2019).

Although the narrow focus on a single active ingredient, such as glyphosate, may have helped to trigger a public debate around food safety and the danger of corporate interests in our food system, a more holistic approach to crop protection and food production is required that takes these externalities into account (IPES-Food, 2016).

Managing weeds is a key challenge for farmers across the globe. While mulching, crop rotation, intercropping and green manures are all options, additional weeding may be required – often by appropriate, small machines. Alternatives to herbicides do exist. For commercial (conventional and organic) farmers, affordable mechanical weeding technologies would make a huge difference. For instance, the food processing industry has benefited a lot from optic food sorting machines. In a fraction of a second, a stone the size of a pea can be removed from millions of peas.

Despite what the industry wants to make us believe, farmers do not need herbicides. If countries are serious about public health, more research is needed to support non-chemical food production.

If governments would invest more in alternatives to chemical agriculture and organize nation-wide campaigns (as they have done for decades to inform people of other health risks such as smoking, and drinking and driving), farmers, gardeners and shopkeepers (like the lady near my village) would become more aware of the dangers of herbicides and more open to promoting and using alternatives.

I walked out of the village garden shop without my organic chicken feed. It was not in stock, due to low demand. I reflected on how shopkeepers are happy to sell what customers demand. It is up to us, the consumers, to lead the way. I hope one day to go back and find this shop selling better tools for controlling weeds, and organic feed for my chickens.*

◼️📹 Related video

Farmers in West Africa show how to control weeds without herbicides.

AfricaRice, Agro-Insight, IER, Intercooperation and Jekassy (2008) *Effective Weed Management in Rice*. Hosted by Access Agriculture. 15 minutes.
www.accessagriculture.org/effective-weed-management-rice

A version of the section above was previously published on www.agroinsight.com in 2019, by Paul Van Mele.

The Times They Are a-Changin'

Talking with my farmer neighbours in Belgium in early 2022, they all wondered how they would manage, as the cost of chemical fertilizers had risen by 500%. They paid 1 Euro per kilogram of fertilizer. The 2022 Russian invasion of Ukraine was having serious consequences on the global food supply, as Russia was the world's top exporter of nitrogen fertilizers and the second leading supplier of both potassic and phosphorous fertilizers. While farmers across the globe are known to be creative and adaptive (Hansson, 2019), the changes required in the near future will be of a scale unseen before. The spike in fuel and fertilizer prices may be a fundamental trigger.

In Belgium, 2018 and 2019 were unusually hot and dry. While some farmers thought it was necessary to start looking into growing other crops, most farmers pumped up more groundwater to irrigate their maize or pasture to feed their animals, even with signs of depleting groundwater reserves, which also affected the vitality and survival of trees.

The COVID-19 crisis during the following 2 years affected global trade and showed everyone in the food sector how dependent we have become on imports, be it for food, feed or materials needed to process and package food. To rely less on soy bean imports, several farmers in Belgium started to grow their own legume fodder like lucerne (alfalfa), a shift promoted by European policies.

While an acute crisis often makes people see things more clearly, some changes in our environment have gone unnoticed by the public for decades; only now alarm bells are starting to go off. The long-term use of agrochemicals in monocrop farming has had a devastating effect on our biodiversity. In the beekeepers' association of my eldest brother Wim in Flanders, in northern Belgium, all members, including those who had spent a lifetime caring for bees, reported that three out of four colonies died in the winter of 2021–2022. Jos Kerkhofs, my beekeeper friend in Erpekom in eastern Belgium where I live, has seen the same trend among the members of his association. While local honey has become a rare commodity this time of the year, the wider consequences on society are seriously worrying (Figs 1.5 and 1.6), as 84% of our crop species rely on or partly rely on honey bees and wild pollinators (Kluser *et al.*, 2010).

A global decline in pollinators is due to a loss of natural habitats, massive use of herbicides and other pesticides, the massive increase in roads and motorized traffic as well as a continent-wide nocturnal light pollution and nitrogen deposition (van Lexmond *et al.*, 2015).

All the above examples are what one refers to as externalities, or external costs. An external cost is one not included in the market price of the goods and services being produced, or a cost that is not borne by those who create it. And as we start to realize, the external costs of climate change, the depletion of natural resources such as groundwater and the loss of biodiversity, will need to be paid by society.

Media plays a big role in sensitizing people about these matters. In Belgium, radio programmes and news items on these matters have become much more common since 2022. A spokesperson from the food industry said that three out of five of the main food companies considered closing their doors. Because of the rising fuel prices and costs of raw materials, they were unable to continue providing food to supermarkets at the same rock-bottom prices. Unless supermarkets show some flexibility and are ready to pay more, food companies will go bankrupt.

Is this the era when cheap food will come to an end because it is no longer feasible to ignore the rising costs of chemical fertilizers? Hopefully, facing up to these costs will lead policy makers, farmers and others in the food system to make drastic decisions and embrace agroecology, home gardening, short food supply chains and less processed food, to care for our environment and for our farmers. The world's farmers can contribute by experimenting with ecological alternatives, like green manures, organic growth promoters, intercropping, cover crops and mulch.*

The Pollinators

Pesticides and other threats have led to a
dramatic decrease in Europe's pollinators

1/10 European bee and butterfly species in Europe are in danger of extinction.

84%
of crop species

73%
EU wild flowers

1/3 of bees and butterflies
are in decline.

rely on or partly rely on insects.

€15 billion of the EU's annual agriculture output is
directly attributed to insect pollinators.

Source: IUCN European red list 2015 and 2019

EURACTIV

Fig. 1.5. Status of pollinators in Europe. Photograph used with permission from Euractiv
Media Network.

Related video

Farmers and others can use simple techniques to provide habitats for pollinators and other
beneficial insects.

Agro-Insight, IRD and INIAF (2022) *Flowering Plants Attract the Insects that Help Us.*
Hosted by Access Agriculture. 15 minutes.
www.accessagriculture.org/flowering-plants-attract-insects-help-us

---

**A version of the section above was previously published on www.agroinsight.com
in 2022, by Paul Van Mele.*

Fig. 1.6. Pollinators are crucial to life on Earth.

European Deserts, Coming Soon

Not a single day passes without news on the radio of the increasing challenges farmers face in Europe and in much of the world. Rainfall has become more erratic and intense, and heatwaves are becoming the new normal. We should all be concerned about the speed at which climate change is affecting our planet. How we decide to live, what to eat, where to source our food and how we spend our leisure time cut across all of society: agriculture, industry, tourism and transport, as well as urban planning and rural land use.

According to the main Spanish farmers' association, the Coordinator of Farmers' and Livestock Owners' Organizations (COAG – *Coordinadora de Organizaciones de Agricultores y Ganaderos*), drought affects 60% of Spain's countryside and has destroyed crops on 3.5 million hectares. This is more than double the farmland we have in Belgium and six times that of Flanders.

Three years of very low rainfall and high temperatures have put Spain officially into long-term drought. In 2023, losses were not limited to wheat, barley and maize, but also affected nuts, orchards, vineyards and olives, as well as sunflower and vegetable farming. As vegetation is scarce, bees are failing to make honey. Beekeepers were facing a third consecutive season without a harvest. According to a CNN article, 'Disappearing lakes, dead crops and trucked-in water', these conditions point to a new reality for parts of Europe, which is warming twice as fast as the global average (CNN, 2023).

Farms are not the only places in trouble. Municipal water systems are dryer across much of southern Europe. In Italy, in April of 2023 extreme drought was already affecting Lombardy and Piedmont, with 19 towns experiencing the highest level of shortage. Some places have already started receiving water in tanker trucks (Bertelli, 2023).

As it takes 15,000 l of water to produce just one kilogram of beef and 3000 l of water to produce a kilogram of cheese, it is no wonder that livestock farming is also at risk. After all, farmers need pasture to feed their animals. When a choice has to be made between ensuring drinking water for its citizens or irrigating pasture and crops to feed animals, governments usually favour cities over farms, but countries need both.

Often governments come up with short-term solutions, such as asking people to stop watering their lawns or washing their cars, instead of planning for long-term measures to conserve water. Different food and fodder crops (and combinations) need to be promoted; more trees and living hedges need to be planted in and around fields to reduce evaporation by the hot sun and dry winds. Above all, measures are needed to store more carbon in the soil.

Enabling citizens to have online access to real-time monitoring of the depth and quality of the groundwater, a natural resource that is usually invisible, could help to sensitize all of us to the need to conserve water. Groundwater levels drop quickly due to continuous pumping and are only partially recharged. Water levels in aquifers around the world continue to decline (Schwartz and Ibaraki, 2011). While monitoring is needed, let us not be naïve and think that when the groundwater table is recharged, households, industry and farming can go back to using water in an unlimited way. Already in rural Belgium, even though we had the wettest spring in 2023 since records began to be kept in 1833, in the countryside we see many old oak trees suffering from the three previous years of drought and heat stress.

In many European countries, agricultural lobby groups aggressively sustain their opinion that the livestock sector should not shrink in order to survive. With all that is happening, how long will they be able to continue taking such an unrealistic position? Animals can be properly fed with technologies that use water and fossil fuels more efficiently, but technical solutions also have their limitations. If lobby groups do not help their members to adapt, the climate will soon dictate what is possible and what is not. The changes we see in southern Europe should set off alarms in the north as well.

If we are unable to quickly and drastically curb greenhouse gas emissions, replenish water resources and invest in an agriculture that helps to cool the planet, rather than deplete its natural resources, we will soon be wondering why many of our favourite food products are no longer available or affordable.*

> 📹 **Related video**
>
> ---
>
> In a drier world, farmers will appreciate water-saving techniques like hydroponic fodder.
>
> Pagar, A. (2018) *Hydroponic Fodder*. Hosted by Access Agriculture. 14 minutes.
> www.accessagriculture.org/hydroponic-fodder

*A version of the section above was previously published on www.agroinsight.com in 2023, by Paul Van Mele.

The Nitrogen Crisis

To reduce the impact of climate change and to nurture resilient food systems, we need to redesign them. Today, our food systems account for nearly one third of global greenhouse gas emissions. In line with the United Nations Environment Programme (UNEP, 2019) Colombo Declaration on Sustainable Nitrogen Management, the European Commission developed its Farm to Fork Strategy, which is at the heart of the European Green Deal to make Europe climate neutral by 2050 (European Commission, 2019). The Farm to Fork Strategy defines targets to reduce nutrient losses by at least 50% by 2030, linked to integrated nutrient management action plans to prevent further deterioration of soil fertility. The use of synthetic nitrogen fertilizers is expected to be reduced by at least 20%.

The societal cost of nitrogen pollution is dominated by the impact of ammonia on human health and of nitrates on water bodies. To achieve the targeted 50% reduction, Leip and colleagues (2022) found that improved on-farm nitrogen management could only reduce nitrogen losses by 37%. A 50% reduction would require using other measures as well, such as eating diets with less meat. Trying to come to a consensus on how farmers should be supported has widened political divides.

As is often the case, looking at history helps us to better plan for the future. From 2005 to 2013, across Europe, the number of farms with less than 50 ha of land steadily decreased, while those between 50 and 100 ha remained more or less stable. Those over 100 ha slightly increased. Yet more than half of the farming population in Europe is older than 55 years (EuroStat, 2022a). Meanwhile, the younger farmers have invested in labour-saving equipment, for example to work the larger holdings, acquiring high debts along the way. Further investment in climate mitigation will require proper support so that when the older farmers retire, the next generation will be able to cope with the ever-increasing pressure of bank loans.

The war in Ukraine has triggered a sharp rise in the price of artificial fertilizers, making chemical-based farming less profitable. It is estimated that globally only one third of the applied nitrogen from chemical fertilizers is used by crops (Anas *et al.*, 2021). In addition to the mounting pressure on farmers to help mitigate climate change by reducing carbon and nitrogen emissions, farmers are keen to optimize the use of animal manure.

While animal and human manure has been used to keep soils fertile for thousands of years, something has gone wrong in the recent past. In a German documentary on the Aztecs, called *Children of the Sun*, ethnologist Antje Gunsenheimer describes some ancient human manure management (Oblaender and Reiß, 2020). The central market in Tenochtitlán, the Aztec capital, had public toilets where urine and faeces were collected separately in clay pots. Dung traders sold the composted dung as fertilizer, while the urine was used for dying fabric and leather tanning.

From the earliest days of farming in Europe, animals were kept on a deep bedding of straw. Such practices survived in Portugal until the 1980s (Bentley, 1992) and are still practised by some organic farmers. But nowadays most animals in Europe are kept on a metal grid, and the mix of urine and dung is collected in large, underground reservoirs. When excrement and urine from cows or pigs mix, a lot of methane gas (CH_4) and ammonia (NH_4) is produced. The old practices of using straw as bedding, as well as innovative designs to separate the dung from the urine, are getting some renewed attention in livestock farming, because when they are separated, greenhouse gas emissions can be reduced (Galama *et al.*, 2020). Bedding with crop residues such as wheat straw may provide substantial benefits.

Engineers in the Netherlands, the USA, Israel and various other countries are researching how best to adjust modern livestock sheds. Some promising examples include free walk housing systems that operate with composting bedding material or artificial permeable floors where cows can move about and lay down. Other sustainable techniques that are being explored include the CowToilet, which separates faeces and urine. As converting housing systems may be costly and therefore only adopted slowly by farmers, it is important to also experiment with better ways of applying liquid manure (Galama *et al.*, 2020).

In modern livestock systems, urine and manure are mixed with the water used to wash the pens. Getting rid of this slurry, or liquid manure, has become a main environmental concern, especially because manure often contains poorly metabolized veterinary products, such as antibiotics, and heavy metals, such as cadmium, zinc and copper from feed additives (Köninger *et al.*, 2021). When liquid manure is applied to the soil, much of the nitrogen evaporates as nitrous oxide or N_2O, a greenhouse gas 300 times more powerful than carbon dioxide. Another fraction is converted to nitrates (NO_3), which seep through the soil and pollute the groundwater. While manure used to be a crucial resource, it has now become a waste product and an expense for farmers.

Making better use of animal waste will be crucial for the future of our food. One key factor is the lack of soil organic matter and good microbes that can help capture nitrogen and release this more slowly to benefit crops.

Solutions that are financially feasible for farmers will require the best of ideas, with inputs from farmers, soil scientists, microbiologists, ecologists and chemical and mechanical engineers, as well as social scientists.

Practices that help to build up soil carbon will be crucial to reduce the environmental impact of animal manure and fertilizers. Ploughing is known to have a detrimental effect on soil organic matter, as it induces oxidation of soil carbon (Montgomery, 2007). Reduced tillage or zero tillage for crop cultivation, and regenerative farming to make animal farming more sustainable, has been promoted and used in the USA and other parts of the world and could be explored more intensively in Europe.

Also, there will be a need to revive soil microorganisms, as these have been seriously affected by the use of agrochemicals and the reduced availability of soil organic matter. The expensive machines that are currently used by service providers to spread or inject liquid manure in farmers' fields could equally be used to inject solutions with good microorganisms that would help to capture nitrogen to then release it to crops, and build up a healthy soil.

Despite the gains that can be achieved, the high-tech option to reduce nitrogen losses by 50% without dietary change is the least desirable strategy (Leip *et al.*, 2022). Human creativity will be required to devise solutions that are economically feasible for farmers and that contribute to more resilient and sustainable food systems. To make this happen, more investments are required in research that truly addresses the fundamentals of the problems. Far too much public money is still invested in research on agrochemicals, herbicide-resistant crop varieties and livestock feed additives, all of which benefit large corporations.*

Related video

Farmers in Bangladesh produce their own good microorganisms, which they use to keep their cattle healthy.

Siddique, R.K. (2022) *Using Good Microorganisms in Cattle Farming*. Hosted by Access Agriculture. 13 minutes.
www.accessagriculture.org/using-good-micro-organisms-cattle-farming

---

A version of the section above was previously published on www.agroinsight.com in 2022, by Paul Van Mele.

Big Chicken, Little Chicken

In her McKenna, 2017 book *Big Chicken*, Maryn McKenna tells the story of antibiotic abuse in agriculture. In the 1940s the US military used antibiotics to treat soldiers suffering from infectious disease, one of the first large-scale uses of these drugs. Penicillin had been placed in the public domain, and the world was in an optimistic mood at having a widely available treatment for common infections.

Soon after the war, in 1948, British-American scientist Thomas Jukes, working at Lederle Laboratories in New Jersey, showed that chickens gained more weight when their food included antibiotics, even on a poor diet. This was a crucial discovery for industrializing chicken rearing. Until then, poultry in the USA were mostly reared in small batches, allowed to range freely in fields, where they scratched a natural diet of plants and bugs, sometimes supplemented with fish meal. Chickens are by nature highly omnivorous. Discovering that antibiotics helped to fatten chickens meant that the birds could be fed cheap, low-grade maize and soy bean.

The FDA (Food and Drug Administration) approved antibiotics as a growth regulator in the USA in 1951. By the 1950s, a successful chicken farmer in Georgia, Jesse Dixon Jewell, had begun to expand his operation by selling chicks and feed to other farmers, and buying their finished birds. The feed was laced with antibiotics, partly to boost growth but also to control bacterial diseases. Industry would soon follow this twin example of farming out the birds to contract growers and including the antibiotics in their pre-pared rations.

By 2001, Americans were taking 3 million pounds (1.4 million kilograms) of antibiotics annually, while US livestock was being dosed with 24.6 million pounds (11.2 million kilograms).

For many years there were few if any concerns about this unprecedented use of a human drug to boost food production, and cheap meat was certainly popular. But this relaxed attitude began to change when research showed that much of the antibiotics ended up in the meat and eggs that consumers ate. This widespread use meant that once valuable drugs began to be compromised as bacteria that caused disease in humans became resist-ant to the antibiotics.

Under pressure from pharmaceutical companies, the US government was slow to restrict antibiotics as animal growth promoters. Finally, the large poultry companies began to self-regulate. By about 2009 they realized that they could produce birds without antibiotics, simply by using vaccines and improving farm hygiene. By 2014 some of the largest producers in the US, like Foster Farms and Perdue Farms, had stopped feeding antibiotics to chickens. Various grocery stores and fast-food chains soon banned chicken raised on antibiotics.

In September 2016 the UN moved to curb non-prophylactic antibiotic use in animals, which was linked to an estimated 700,000 human deaths worldwide, per year. In North America and Western Europe antibiotic abuse was by then largely solved, thanks to improved industry standards, govern-ment regulations and public awareness. But McKenna cautions that livestock antibiotic abuse remains a worrying problem in much of South America, South Asia and China.

After I finished reading *Big Chicken*, my wife Ana and I visited *La Cancha*, the vast, open-air market that still functions in Cochabamba, where we bought a little grey hen and a big red one. Feeding the hens was a chance to learn what smallholder farmers have always known: that chickens are as omnivorous as people. Ours preferred the smaller, denser grains like sorghum to maize. Chickens especially like sow bugs (woodlice), the little roly-poly crustaceans (order Isopoda) that live in leaf litter worldwide. Our hens learned to knock the seeds off of amaranth plants and then eat the seeds from the ground. Chickens also like table scraps. The longer we keep these birds, now named Oxford and Cambridge, the bigger their eggs get. They both lay an egg every day; clearly the hunter-gatherer diet agrees with them.

The problem, as McKenna explains, is that factory farming made chicken as cheap as bread in the USA and Europe. People living in low-income countries now want their chance at cheap meat. Chicken is cheap in Bolivia. In the open-air market, it sells for just 10 Bolivianos (US$1.40) per kilogram and fried chicken restaurants have sprung up all over the city.

Rearing chickens has become a new industry in Bolivia. Farmers can make a barn with cheap lumber and plastic sheeting, buy the day-old chicks and purchase the feed by the bag or the tonne. No doubt many of the poultry producers in Bolivia are careful and conscientious. But other growers raise their birds on feed blended with antibiotics, labelled as a growth promoter, and there is little public awareness of the risk of antibiotics in animal feed. While there are compelling reasons to reduce the cost of food in low-income countries, the Global South also needs to consider the risks of animal antibiotics.*

◼ Related video

In this video from India, farmers explain how to care for chickens in simple, natural ways.

Pagar, A. and Anthra (2021) *Natural Ways to Keep Chickens Healthy*. Hosted by Access Agriculture. 9 minutes.
www.accessagriculture.org/natural-ways-keep-chickens-healthy

A version of the section above was previously published on www.agroinsight.com in 2018, by Jeff Bentley.

References

Anas, M., Liao, F., Verma, K.K., Sarwar, M.A., Mahmood, A. *et al.* (2021) Fate of nitrogen in agriculture and environment: Agronomic, eco-physiological and molecular approaches to improve nitrogen use efficiency. *Biological Research* 53, 47.

Bentley, J.W. (1992) *Today There Is No Misery: The Ethnography of Farming in Northwest Portugal.* University of Arizona Press, Tucson, Arizona.

Bertelli, M. (2023) 'No water, no life': Drought threatens farmers and food in Italy. Available at: www.context.news/climate-risks/no-water-no-life-drought-threatens-farmers-and-food-in-italy (accessed 16 October 2024).

Bonifacio, A., Aroni, G., Villca, M. and Bentley, J.W. (2022) Recovering from quinoa: Regenerative agricultural research in Bolivia. *Journal of Crop Improvement* 37, 687–708.

Brown, G. (2018) *Dirt to Soil: One Family's Journey into Regenerative Agriculture.* Chelsea Green Publishing, White River Junction, Vermont.

Cansler, C. (2013) Where's the beef? *Distillations Magazine*. Science History Institute Museum and Library. Available at: www.sciencehistory.org/distillations/wheres-the-beef (accessed 16 October 2024).

CNN (2023) Disappearing lakes, dead crops and trucked-in water: Drought-stricken Spain is running dry. Available at: https://edition.cnn.com/2023/05/02/europe/spain-drought -catalonia-heat-wave-climate-intl/index.html (accessed 16 October 2024).

Defarge, N., Spiroux de Vendômois, J. and Séralini, G.E. (2018) Toxicity of formulants and heavy metals in glyphosate-based herbicides and other pesticides. *Toxicology Reports* 5, 156–163.

European Commission (2019) The European Green Deal. Available at: https://ec.europa.eu/ info/strategy/priorities-2019-2024/european-green-deal_en (accessed 16 October 2024).

European Commission (2021) Communication from the Commission to the European Parliament, the Council, the European Economic and Social Committee and the Committee of the Regions on an Action Plan for the Development of Organic Production. Available at: https://ec.europa.eu/info/food-farming-fisheries/farming/organic-farming/organic-action-plan_en (accessed 16 October 2024).

EuroStat (2022a) Farmers and the Agricultural Labour Force – Statistics. Available at: https:// ec.europa.eu/eurostat/statistics-explained/index.php?oldid=431368#Farm_managers_ are_typically_male_and_relatively_old (accessed 16 October 2024).

Galama, P.J., Ouweltjes, W., Endres, M.I., Sprecher, J.R., Leso, L. *et al.* (2020) Symposium review: Future of housing for dairy cattle. *Journal of Dairy Science* 103, 5759–5772.

Hansson, S.O. (2019) Farmers' experiments and scientific methodology. *European Journal for Philosophy of Science* 9, 1–23.

Harper, D. (2001) *Changing Works: Visions of a Lost Agriculture*. University of Chicago Press, Chicago, Illinois.

HLPE (2019) Agroecological and Other Innovative Approaches for Sustainable Agriculture and Food Systems that Enhance Food Security and Nutrition. Report by The High Level Panel of Experts on Food Security and Nutrition (HLPE), FAO, Rome, Italy. Available at: www.fao.org/fileadmin/user_upload/hlpe/hlpe_documents/Hlpe_Reports/Hlpe-Report -14_EN.pdf (accessed 16 October 2024).

IPES-Food (2016) *From Uniformity to Diversity: A Paradigm Shift from Industrial Agriculture to Diversified Agroecological Systems*. International Panel of Experts on Sustainable Food Systems. Available at: https://ipes-food.org/report/from-uniformity-to-diversity/ (accessed 16 October 2024).

Kluser, S., Neumann, P., Chauzat, M.P., Pettis, J.S., Peduzzi, P. *et al.* (2010) Global honey bee colony disorder and other threats to insect pollinators. UNEP Emerging Issues 1-16. Available at: https://i.unu.edu/media/ourworld.unu.edu-en/article/3163/UNEP-Emerging-Issues-bees.pdf (accessed 16 October 2024).

Köninger, J., Lugato, E., Panagos, P., Kochupillai, M., Orgiazzi, A. *et al.* (2021) Manure management and soil biodiversity: Towards more sustainable food systems in the EU. *Agricultural Systems* 194, 103251.

Leip, A., Caldeira, C., Corrado, S., Hutchings, N.J., Lesschen, J.P. *et al.* (2022) Halving nitrogen waste in the European Union food systems requires both dietary shifts and farm level actions. *Global Food Security* 45, 100648.

McKenna, M. (2017) *Big Chicken: The Incredible Story of How Antibiotics Created Modern Agriculture and Changed the Way the World Eats*. National Geographic, Washington, DC.

Montgomery, D.R. (2007) *Dirt: The Erosion of Civilizations*. University of California Press, Berkeley, California.

Montgomery, D.R. (2017) *Growing a Revolution: Bringing Our Soils Back to Life*. Norton, New York.

Oblaender, C. and Reiß, A. (2020) *'Children of the Sun' The Aztecs*. TV episode.

Phys.Org (2023) Glyphosate: Where Is It Banned or Restricted? Available at: https://phys.org/news/2023-09-glyphosate-restricted.html (accessed 16 October 2024).

Pimentel, D.C., Harvey, C., Resosudarmo, I., Sinclair, K., Kurz, D. *et al.* (1995) Environmental and economic cost of soil erosion and conservation benefits. *Science* 267, 1117–1123.

Pollan, M. (2008) *In Defense of Food: An Eater's Manifesto*. Penguin Books, London.

Schwartz, F.W. and Ibaraki, M. (2011) Groundwater: A resource in decline. *Elements* 7, 175–179.

UNEP (2019) Colombo Declaration on Sustainable Nitrogen Management. Available at: https://www.inms.international/sites/inms.international/files/Colombo%20Declaration_Final.pdf (accessed 16 October 2024).

van Lexmond, M.B., Bonmatin, J.-M., Goulson, D. and Noome, D.A. (2015) Worldwide integrated assessment on systemic pesticides: Global collapse of the entomofauna, exploring the role of systemic insecticides. *Environmental Science and Pollution Research* 22, 1–4.

2 Eroding Food Cultures

Abstract

Traditional food cultures are eroding due to industrialization, globalization and shifting lifestyles. Case studies from diverse regions, including the French Alps, South Korea and Bolivia, illustrate the decline of self-sufficient farming systems and the associated loss of biodiversity and local knowledge. The negative impacts of processed foods and excessive pesticide use on human health and traditional diets are also evident. Revitalizing local food cultures is essential for the well-being of individuals and communities, requiring a move towards more sustainable and resilient food systems. This revitalization involves preserving traditional knowledge, supporting local farmers and food processors, and promoting conscious consumer choices.

A Lost Alpine Agriculture

Much of local food culture is rooted in local farming traditions. In the mid-1980s, an American anthropologist, Brien Meilleur, studied farming in Les Allues, a village in the French Alps. Meilleur was especially well qualified for the topic, as decades earlier, his own father had left Les Allues for the USA (Meilleur, 1986).

Meilleur interviewed elderly farmers at length about the days of their youth, as far back as the 1940s. These retired farmers explained how families worked in synchrony with nature. They had large cereal fields, divided into many individual plots. Each year they agreed upon a time to plough, and each household would plough their own small plot within the big field. By ploughing and planting at the same time they avoided trampling each other's grain crop. The big fields were on a 3-year rotation, beginning with rye, then barley and finally fallow-plus-pulses. This was similar to the open field system, a functional arrangement which persisted in England until broken up by the enclosure movement after 1773 (McCloskey, 1975).

French Alpine farmers made wine and hard apple cider from fruit they grew themselves. Each family had apple trees near their homes. The village had a special vineyard, at the lowest elevation in the community, where each household had a small plot of precious vines. The villagers wintered cows, sheep and goats in stables, moving them in the spring to montagnettes, cabins above the hamlets where the families made their own hard, low-fat Alpine Tomme cheese. Then every year on 11 June, in a grand procession, the whole village would move their livestock to the high Alpine pastures, with cowbells ringing and dogs barking. The animals would graze communally, on named pastures, moving uphill as summer progressed, until the cows were brought back down on 14 September. Outside specialists were hired to come to turn the milk into cheese, mostly a fine gruyere, which the locals sold (Meilleur, 1986).

Fruits and vegetables were prized, but difficult to come by. The women would tend small household gardens in the spring, as soon as the danger of frost had passed. They produced their own vegetable seed. The families kept bees, which foraged among the pear and apple trees, special varieties that bore fruit which could be stored over the winter.

Barnyard manure provided all the fertilizer the farms needed. To save on firewood, neighbours baked their bread on the same day in ovens in the hamlet square. About 80–90% of what people ate came from Les Allues itself. The roots of this rural economy went back to at least the 1300s, if not earlier. But, as Meilleur explains, this farming system had collapsed by about 1950, at least in Les Allues. He mourns the loss of this way of life, and as I read his moving account, I couldn't help but share in his sadness.

The collapse came about in part because of emigration. Young people were leaving Les Allues for the cities as early as the 19th century. But there were other reasons for abandoning agriculture. After the Second World War, the villagers sold much of their farmland to the Méribel Ski Resort, established just above the highest of the village's hamlets. There were now lots of jobs for local people, on the ski slopes and in the busy hotels, shops and restaurants. The vacationers even visited the beautiful village in the summer, for golf, tennis and mountain biking, so there was employment year-round. The youth of Les Allues no longer had to leave home to find work; the jobs had come to them (Meilleur, 1986).

The old agricultural landscape changed quickly, as the pastures became ski trails and the fields grew wild with brush. The livestock were sold off and the apple trees were strangled by mistletoe, as people abandoned a way of living that was (in today's jargon) sustainable and carbon neutral, and the bedrock of their community.

It is easy to romanticize a healthy rural lifestyle, but the good old days had some rough times too. The farmers of Les Allues managed erosion in their cereal fields by hand-carrying the earth from the bottom furrow to the top of the field every year, the most back-breaking soil conservation method I've ever heard of. For 6 weeks in July and August, people cut hay for 6 days a week from 5 AM to 10 PM, to feed their animals over the winter. To save on fuel, the families would spend winter evenings sitting in the barn, where the cows gave off enough heat to keep everyone warm. People ate meat once a week, maybe twice.

Given the amount of hard work, and the low pay, it is understandable that the young people of Les Allues left farming. It happened all over Europe (Bade, 2008). The people of Les Allues were lucky: young people who wanted to stay in their home villages could work in the ski industry.

As food culture adapts to urbanization, local cuisines keep some old favourites. It would be an understatement to say that you can still get a nice assortment of wines and cheeses in France. As the British biologist Collin Tudge (2004) wrote, all of the world's great cooking is based on traditional farming.*

◼️ Related video

Rural lifestyles form the basis of traditional foods, whether in the Alps or along the Nile.

Nawaya (2017) *Making Pressed Dates*. Hosted by Access Agriculture. 12 minutes.
www.accessagriculture.org/making-pressed-dates

A version of the section above was previously published on www.agroinsight.com in 2021, by Jeff Bentley.

Formerly Known as Food

In a 2018 book, *Formerly Known as Food*, Kristin Lawless cautions readers about the risks of eating processed food produced by industrial farming. For example, maize and soy beans are widely used in animal feeds and edible oils. In the USA, maize and soy beans have been genetically modified to withstand massive applications of glyphosate herbicide. Glyphosate is reported by the World Health Organization (Bergman *et al.*, 2012) to be an endocrine-disrupting chemical (EDC). Although more research is needed to show how these chemicals impact our health, an EDC interferes with the normal working of hormones, the chemical messengers of our bodies. Glyphosate is just one of many chemicals for which health concerns are mounting (Lawless, 2018).

A chemical used in plastic packaging, BPA (bisphenol A), has recently been classified as an EDC, and it is slowly being removed, albeit on a voluntary basis (ChemSec, 2019). BPA is found in everything from plastic milk jugs to the linings of the cans of tinned food, where the BPA leaches into the food. Some companies now offer plastics made from BPS (bisphenol S), cynically advertised as 'BPA free', even though BPS is similar to BPA and is also an EDC (Lawless, 2018).

While some industrial foods are tainted by chemicals, other food products are a health risk in their own right, because of what has been removed from them. For example, industrial vegetable oils, shortenings and margarine have been heated to such high temperatures that their naturally occurring molecules have been broken down and oxidized, their nutritional properties diminished. These factory-made oils are often advertised as 'heart safe', but they actually damage the walls of one's arteries (Lawless, 2018).

Lawless also offers valuable suggestions for healthier eating. For example, cook at home, eat less fast food and skip processed food. Eat whole foods, like whole milk, and real eggs. She advocates joining a food cooperative that works with concerned family farmers who provide healthy food that goes beyond organic. Consumers are not just passive victims of harmful food systems. When we buy ecologically sound and socially responsible food, we use our food money to proactively support change in the right direction.*

◼️ Related video

There are many ways that farmers can manage their harvest ecologically, for a safe product.

AMEDD (2017) *Storing Fresh and Dried Tomatoes*. Hosted by Access Agriculture. 13 minutes.
www.accessagriculture.org/storing-fresh-and-dried-tomatoes

---

*A version of the section above was previously published on www.agroinsight.com in 2019, by Jeff Bentley.

Korean Food Culture

Much of this book is about food: how it is produced, processed and marketed. During a visit with my wife, Marcella, to South Korea, where I was invited to give a talk at the 5th Organic Asia Congress in October 2022, we discovered little by little what makes Korean food culture stand out from other places we have visited over the years.

Many of the local dishes contain a dozen or so little bowls and plates, each with a different salted and fermented vegetable, called kimchi. These are commonly combined with a bowl of sticky rice and a bowl of tasty seaweed soup. A healthy food, kimchi is rich in vitamin C, beta-carotene and other antioxidants, as well as lactic acid bacteria which are good for your guts.

Traditionally, kimchi was prepared in large earthenware pots at the start of autumn to ensure there was enough healthy food during the winter and early spring. To avoid the pots getting damaged by severe frosts, they were buried in the soil, which was said to add a nice flavour to the vegetables (Fig. 2.1).

Making kimchi used to be an annual, communal activity. Neighbouring women would rotate their schedules to help each other prepare large quantities. Each household had its own recipes, while the taste was also influenced by the region. Nowadays, as most people live in cities, few still prepare their own kimchi. People buy different types of kimchi in closed glass jars at the market or in shops, and now have a separate fridge to store them in.

Fig. 2.1. Traditional Korean food culture, whereby rural women help each other to ferment vegetables in large earthen pots, is rapidly disappearing due to urbanization and modern lifestyles.

During the first week of our stay, our host, Jennifer Chang, explained the origin of another dish, called *bibimbap*. To eat it, you need to transfer your rice into a large bowl with a mix of fermented and steamed vegetables, and stir it until all the ingredients are properly mixed. While in the past this was a way to make use of all leftover food, without letting anything go to waste, it now has become like a national dish.

One of the interesting things to do when travelling is to visit a fresh market, as this often gives you a quick glimpse into local food culture. What struck us as unique when visiting Seongdong market in Gyeongju City was the sheer number of microenterprises that processed food.

Soon after entering the covered, daily market, we were drawn to the tantalizing smell of roasted sesame seed. When we reached the shop, we saw that a lady in her fifties was operating various processing units. In one, sesame was roasted over a gas fire while being slowly, mechanically stirred. Next to it, a small machine extracted sesame oil, another popular ingredient of many of the local dishes. Making optimal use of the small space in her shop, the woman used yet another machine to process chilli into flakes. It was a mild variety of chilli that is a steady ingredient in making kimchi.

Another shop had small, stainless steel food processing equipment to make pastries, while yet another one was all set up to extract and pasteurize juices.

While raw fish is common in most fresh markets across Asia, it was nice to see how a few entrepreneurial women were adding value to seafood. To cater for the demanding city dwellers, appealing dishes were being prepared on the spot with a diversity of shrimps, mussels, cockles, squid and fish.

Covered with a transparent foil, these artistically arranged dishes were a real pleasure to the eye.

One place in the market was devoted to serving prepared food (Fig. 2.2). When we finally decided from which buffet to fill our plates, it struck me how local people carefully arranged the various ingredients, either by colour or some other way. Everybody's plate looked really appealing, almost like a painting, where each colour was added with careful thought.

Local food culture is meaningful in Korea, even as it is being eroded by fast-food chains and supermarkets. In response, the local authorities of Gyeongju City decided to renovate the fresh market and make it more attractive to the local community and foreign visitors. Even strong food cultures need occasional support from local authorities and continued appreciation by the new generation in order to survive.*

🎥 **Related video**

Traditional foodways in many parts of the world include rearing or gathering insects.

Nalunga, J. (2023) *Rearing Crickets for Food and Feed*. Hosted by Access Agriculture. 14 minutes.
www.accessagriculture.org/rearing-crickets-food-and-feed

A version of the section above was previously published on www.agroinsight.com in 2022, by Paul Van Mele.

Fig. 2.2. Making traditional food attractive at the Seongdong market to counter the rising attraction of fast-food chains and supermarkets.

Don't Eat the Peels

Local food cultures are often under threat from rapid technical change, including the abuse of insecticides. I was in Colomi, a potato-growing municipality in the Bolivian Andes, making a video with Paul Van Mele and Marcella Vrolijks, and with Juan Almanza from the Foundation for the Promotion and Research of Andean Products (PROINPA – *Fundación para la Promoción e Investigación de Productos Andinos*). Local farmer don Isidro was kindly helping us find tuber moths to star in the video. Don Isidro was an expert at spotting this serious potato pest, instantly recognizing the entrance tunnels of the tiny larvae in the skin of the discarded potatoes. In the field, he easily saw the difference between the frost-damaged potato plants and those that were wilted because the moth's larvae had tunnelled through the centre of the stalk.

As don Isidro and I worked our way across his field, he casually remarked that he no longer had problems with the Andean potato weevil (*Premnotrypes* spp. and other Curculionidae), once the nemesis of Bolivian farmers. Don Isidro explained that now, farmers simply douse the potato plants with lots of insecticide. While don Isidro was happy to be rid of the weevil, he added offhand that when his family fed the potato peelings to their guinea pigs (kept for meat), the little animals died.

'Aren't you afraid to eat those potatoes yourselves?' I asked.

'No', don Isidro said. 'We don't eat the peels.'

'But don't you think that the whole potato might be poisoned by the insecticide?'

'No, I asked the *ingeniero* (the agrochemical dealer) and he said it was fine to eat the potatoes', don Isidro said.

This stunned me, since I have been eating Bolivian potatoes for a long time. At my house, we often boil potatoes in the peel. Now I had just learned that at least some farmers are producing potatoes so poisonous that they will kill that quintessential lab animal, the guinea pig.

The guinea pig and the potato are both native to the Andes. For centuries they have evolved together, on the farm and on the plate. Yet, crops often coevolve with their pests, like the Andean potato weevil. This flightless weevil became a serious pest when farmers intensified potato cultivation in the second half of the 20th century (Ortiz, 2006). Extensionists promoted crop rotation and other ecological pest control methods (Bentley *et al.*, 2009). Unfortunately, the heavy use of insecticides remained popular.

Government policy in most countries lets the public buy dangerous insecticides over the counter. Few agrochemical dealers have training in agriculture and often misadvise farmers (Struelens *et al.*, 2022). Rapid technical change, especially the wrong kind, can endanger local food cultures and form a threat to public health, especially in countries where food safety regulations and control are not taken seriously.*

Related video

Alternatives to chemical insecticides include herbal pest repellents.

Green Adjuvants (2022) *Herbal Pest Repellent*. Hosted by Access Agriculture. 15 minutes. www.accessagriculture.org/herbal-pest-repellent

*A version of the section above was previously published on www.agroinsight.com in 2023, by Jeff Bentley.

How to Feed Babies

Producing more food does not always mean that people eat better. And the most vulnerable people are babies. A 2012 study by Cornell University nutritionist Andy Jones and colleagues, in northern Potosí, Bolivia, found that boosting farm production increased the workload of young mothers. As women worked harder to grow more potatoes for market, they did not always have the time to feed their babies often enough. The toddlers can suffer if their mothers are working too hard and too long (Jones *et al.*, 2012).

One of Jones's colleagues in that study was an experienced and perceptive Bolivian nutritionist named Yesmina Cruz. She explains that in this part of the Andes, some local beliefs were harmful for babies. For example, mothers believed that if the babies went without food when they were small, they would grow up to be able to withstand hunger when they were big. So, the mothers would avoid giving the breast to their newborns for several days, until after losing the colostrum, the rich, yellowish milk that should be a baby's first, nutritious meal. The mothers did not feed their babies often enough and would often start them too soon on supplementary foods like soups or mush (Cruz Agudo *et al.*, 2010). Younger mothers are changing how they bring up their children, but some of the old ideas persist.

In 2016, I had a chance to work with Yesmina, as she wrote a factsheet on mother's milk as part of a course I was giving. The first day of the course, Yesmina outlined her main suggestions: start breastfeeding on the day the baby is born, give mother's milk (and nothing else) for the first 6 months and keep breastfeeding the baby for at least the first 2 years.

Yesmina tried to share her factsheet with men. They were friendly, but not very interested in breastfeeding, which they saw as women's business. Yesmina had better luck in the village of Phinkina, near Anzaldo, Cochabamba, where she met with three older mothers whose children were grown. Yesmina explained that the main sign of malnutrition was that the babies were small for their age (something a mother may not always realize, especially when malnutrition is widespread). Malnutrition in toddlers can be easily avoided by proper breastfeeding. To Yesmina's surprise, the women didn't think it was a problem if their children were smaller than expected in their early years. 'They can eat when they are youths', one of the women explained (although, in fact, children never fully recover from poor development in early years).

On the other hand, the mothers were obsessed with school. They wanted their kids to do well in school and to finish it. Yesmina realized that talking about school could be a way to get moms, and dads, interested in milk for babies, by explaining that mother's milk helps children grow healthier minds and bodies, so they can do better in school.

By the end of the week, Yesmina had developed six key ideas:

1. Eat well when you are pregnant. Men must be motivated, to encourage their wives, daughters and daughters-in-law to eat better food during pregnancy, and to help them ease up on their workload.
2. Start breastfeeding as soon as the baby is born.
3. Give the baby breast milk (and no other foods) until 6 months of age.
4. Introduce supplementary feeding at 6 months .
5. Continue breastfeeding until the baby is at least 2 years old.
6. Colostrum is the first food that feeds the baby's brain. Babies who are well fed on breast milk will grow up to be children who perform better in school.

Feeding children is not instinctive. It is cultural, and many societies develop unhealthy eating habits (Judiann McNulty, personal communication, Guatemala). Northern countries also have their fair share of food pathologies, such as feeding children sugar and white flour, causing tooth decay that has to be remedied in the dentist's chair. But it is another issue if young mothers know about the importance of breastfeeding but cannot practise it because excessive farm work gets in the way. As Yesmina Cruz has also explained to me, it is not enough to teach young mothers how to feed babies and toddlers if their family is unsupportive. Other household members, especially husbands and mothers-in-law, need to be supportive when young mothers spend more time feeding their little ones.*

📹 Related video

This excellent video from Mali shows how to care for mothers who have recently given birth.

AMEDD (2016) *Helping Women Recover after Childbirth*. Hosted by Access Agriculture. 13 minutes.
www.accessagriculture.org/helping-women-recover-after-childbirth

---

A version of the section above was previously published on www.agroinsight.com in 2016, by Jeff Bentley.

References

Bade, K. (2008) *Migration in European History*. Wiley, Hoboken, New Jersey.

Bentley, J.W., Boa, E., Danielsen, S., Franco, P., Antezana, O. *et al.* (2009) Plant health clinics in Bolivia 2000–2009: Operations and preliminary results. *Food Security* 1, 371–386.

Bergman, A., Bergman, A., Heindel, J.J., Jobling, S., Kidd, K.A. *et al.* (2012) World Health Organization (WHO). Available at: www.who.int/publications/i/item/9789241505031 (accessed 16 October 2024).

ChemSec (2019) Recognition of BPA as an EDC for Human Health Will Increase the Protection of Consumers. International Chemical Secretariat. Available at: https://chemsec.org/recognition-of-bpa-as-an-edc-for-human-health-will-increase-the-protection-of-consumers/ (accessed 16 October 2024).

Cruz Agudo, Y., Jones, A.D., Berti, P.R. and Larrea Macías, S. (2010) Lactancia materna, alimentación complementaria y malnutrición infantil en los Andes de Bolivia (Breastfeeding, supplemental feeding and childhood malnutrition in the Andes of Bolivia). *Archivos Latinoamericanos de Nutrición* 60, 7–14.

Jones, A.D., Cruz Agudo, Y., Galway, L., Bentley, J.W. and Pinstrup-Andersen, P. (2012) Heavy agricultural workloads and low crop diversity are strong barriers to improving child feeding practices in the Bolivian Andes. *Social Science & Medicine* 75, 1673–1684.

Lawless, K. (2018) *Formerly Known as Food: How the Industrial Food System Is Changing Our Minds, Bodies, and Culture*. St. Martin's Press, New York.

McCloskey, D.N. (1975) The persistence of English open fields. In: Parker, W. and Jones, E.L. (eds) *European Peasants and Their Markets: Essays in Agrarian Economic History*. Princeton University Press, Princeton, New Jersey, pp. 92–120.

Meilleur, B.A. (1986) Alluetain ethnoecology and traditional economy: The procurement and production of plant resources in the northern French Alps. PhD dissertation, University of Washington, Seattle, Washington.

Ortiz, O. (2006) Evolution of agricultural extension and information dissemination in Peru: An historical perspective focusing on potato-related pest control. *Agriculture and Human Values* 23, 477–489.

Struelens, Q.F., Rivera, M., Alem Zabalaga, M., Ccanto, R., Quispe Tarqui, R. *et al.* (2022) Pesticide misuse among small Andean farmers stems from pervasive misinformation by retailers. *PLOS Sustainability and Transformation* 1, e0000017.

Tudge, C. (2004) *So Shall We Reap*. Penguin Books, London.

3   Concentration of Power in the Food System

Abstract

The concentration of power in the food system and agriculture's dependence on expensive farm machinery and external inputs have induced falling profits for family farmers, and a disconnect between consumers and producers. Industrial agriculture depletes soil health and reduces biodiversity, while yields have stagnated or declined. Agroecology promotes diversified cropping systems and reduces the dependency on fossil fuel-based machinery and other technologies. Ethical labels like Fairtrade empower farmers and support rural communities. Short food chains reduce our ecological footprint while connecting producers and consumers, but large wholesalers often remove farmer information from packaging, weakening consumer connection to food and limiting farmers' negotiating power.

Stuck in the Middle

The 2021–2023 energy crisis has made chemical fertilizers unaffordable to many farmers, especially in developing countries. Modern agriculture will need to become less dependent on expensive external inputs such as animal feed and synthetic fertilizer, and make better use of knowledge of the ecological processes that shape the interplay between soil, nutrients, microorganisms and plants. But whether farming will remain a viable business for European farmers in the next decade will not only depend on new knowledge.

Since 1980, two thirds of the farmers in Belgium have abandoned this profession, with currently only some 35,000 farmers remaining in business (Statbel, 2023), and many see a bleak future. Large corporations and supermarkets are keeping the price of commodities at rock bottom, sometimes even below the production cost, according to my farmer friends in Belgium. So it comes as no surprise that few young people still see a future in farming. Margins are so slim that a neighbouring dairy farmer in Belgium told me once that the difference of 1 Euro cent per litre of milk he sells can make or

© Jeffery W. Bentley and Paul Van Mele 2025. *Agroecology in Practice: From Local Initiatives to Global Scaling Through Video* (J.W. Bentley and P. Van Mele) DOI: 10.1079/9781800628793.0003

break his year. In 2016, around 30% of French farmers had an income below 350 Euros per month, less than one third of the minimum wage.

On average, one French farmer (often a dairy or beef cattle farmer) commits suicide every day, although in reality this may be higher as figures exclude farmers who might have killed themselves during the year after they stopped farming (Bossard *et al.*, 2016). The suicide rate among Swiss farmers is almost 40% higher than the average for men in rural areas, with the gap increasing since 2006 (Steck *et al.*, 2020). The reasons include financial worries and inheritance problems related to passing the farm on to their children.

How did things get so bad for farmers, and is there still time to change the tide? While reading a book on the history of the Belgian farmers' organization called the *Boerenbond* (Farmers' League), I was struck by how deeply engrained our food crisis is and how history has greatly shaped our agricultural landscape and food crisis (Belgische Boerenbond, 1990).

As the steam engine made it possible to transport food much faster and over longer distances, from 1880 onwards large amounts of cheap food from America, Canada, Russia, India and Australia flooded the European markets. This resulted in a sharp drop in food prices and many farmers were forced to stop or expand, while others migrated to Canada, the USA, Argentina and Brazil.

From the early 1890s, Belgian farmers began organizing into a cooperative to make group purchases of chemical fertilizers, seed, animal fodder, milking machines and other equipment. Milk adulteration was one dubious strategy some farmers used to make a living.

As early as 1902 the *Boerenbond* started providing administrative support to its members. Basically, consultants were recruited, subsidized by the Ministry of Agriculture, to keep an eye on the financial books of farmers, and on the quality of their milk. The Ministry also invested in mobile milking schools to teach farm women about dairy and milk processing. Along with milking competitions, this boosted the attention to quality and hygiene.

The *Boerenbond* increasingly tried to bring various regional farmer organizations and milk cooperatives under its wing. In between the two World Wars they had representatives in Parliament, and they had their own oil mills, warehouses, laboratories and animal feed factory (made, for instance, from waste chaff from the flax industry). The *Boerenbond* didn't risk manufacturing their own chemical fertilizer, but bought shares in some of the large chemical companies. Group marketing, education, social security, credit and insurance were all managed in-house to support its members.

It all seemed so progressive, but by the 1930s, deepened by the stock market crash in 1929, the organization was in a dire financial situation. After the crash of potato and milk prices in 1936, the government realized that the *Boerenbond* was no longer capable of providing all these services, so it set up its own credit and marketing institutions for milk, grain and horticultural crops.

Shortly after the Second World War, the Marshall Plan provided food aid and helped to reconstruct Europe, under the condition that Western Europe subscribe to international free trade. While economic cooperation and integration gradually took shape, the economic advisors of the *Boerenbond* pleaded to keep a certain level of national autonomy for matters related to agriculture. But as food and milk production increased, the need for export markets grew and the *Boerenbond* became a strong advocate of European integration.

In 1958, a year after the European Economic Community was established, member countries developed an agricultural policy meant to guarantee a decent income for farmers. Throughout the 1960s and 1970s, productivity enhancement was considered a priority, but farmers found it hard to keep on investing in restructuring their farms to ever more specialized production units while overproduction resulted in falling prices. In reality, farmers had to take larger loans and earned less and less. As in the USA, European farmers were buying more machinery, paying more for inputs and falling deeper into debt.

In 1984, the European Community introduced production quotas to address the shocking situation of milk lakes and butter mountains. With very narrow profit margins set by a limited number of buyers, many farmers gave up (Belgische Boerenbond, 1990). For those who remained in business, the quotas lasted for about 30 years. By 2015, dairy farmers again could produce as much as they wanted. The European Commission thought that this liberalization would not bring back those lakes and mountains, because there was a growing market from developing countries, including China, and price monitoring had improved. In reality, in an attempt to prop up prices and curb the dairy crisis, Brussels has been buying up milk since 2015. Stockpiled in warehouses, mainly in France, Germany and Belgium, the sacks of milk powder are reminiscent of the milk lakes. Milk farmers and traders fear that these stockpiles are dragging down prices, as buyers expect the dried milk lakes to be sold off at any time (Livingstone, 2018).

Classical economics is based on the idea of many willing buyers and many willing sellers. In modern Europe there are many regulated farmers, buying agrochemicals, seed and animal feed from a few corporations and selling to just a few buyers. Farmers are forced to take prices for inputs set by large corporations, while prices of raw milk are fixed by supermarkets who have concentrated the power of the market. Whether they buy or sell, farmers are price takers, caught in the middle between monopolistic suppliers and a few powerful buyers. And farmers are paying a high price: due to multiple crises, input costs have risen sharply since 2018. The Russian invasion of Ukraine in 2022 has further disturbed global agricultural markets. From 2021 to 2022, the average price of fertilizers in the European Union (EU) doubled (+101%), with sharp rises also for energy and lubricants (+60%) and for animal feed (+35%) (Eurostat, 2022b).

While EU policies can contribute to protecting our farmers and our environment, consumers also have a crucial role to play. As consumers, we have no idea how the continuous search for the cheapest products is putting farmers in a stranglehold. While Fairtrade schemes are a nice thought, in reality all food sold anywhere should be fair for the people who produce it, including our own dairy farmers (IPES-Food, 2019).

For more than a century, strong farmer organizations such as the *Boerenbond* have tried to protect farmers' interests by promoting a model of industrial agriculture. How the *Boerenbond* will deal with farmers' hard realities, the complexities of a changing climate, environmental degradation and economic pressure of corporations and supermarkets will determine its future relevance.

Improved consumer awareness to buy local produce at a fair price, enhanced access to affordable, sustainable animal feed and policies conducive to environmentally sound family farming will decide whether farmers will be able to survive or be replaced by new smart agriculture that can do without farmers, using machineries, chemicals and investment funds.*

▇ Related video

One of a dozen videos available from Access Agriculture for small-scale dairy farmers in developing countries.

Agro-Insight (2016) *Pure Milk is Good Milk*. Hosted by Access Agriculture. 8 minutes. www.accessagriculture.org/pure-milk-good-milk

---

*A version of the section above was previously published on www.agroinsight.com in 2019, by Paul Van Mele.

Hybrid Maize and Chemical Fertilizer Fail to End Poverty

In 2005, Jeffrey Sachs, macroeconomist at Columbia University, started the Millennium Villages Project. At 14 sites across Africa, the project intended to end poverty, to pull people above a daily income of US$1.25 a day, by investing in health, education and agriculture. Sachs started the first 5-year phase of the project with almost US$120 million from a handful of wealthy donors.

As told in Nina Munk's 2013 book *The Idealist*, Sachs was intensely optimistic and sincere. The funding would allow him to try a model to end poverty; he hoped that after some initial success, governments and international agencies would follow with larger investments to end poverty worldwide (Munk, 2013).

The villages were actually towns, with an average of about 6000 residents. In each one, the project was led by an educated local person who shared Sachs's vision.

Journalist Nina Munk followed Sachs for 6 years, and also visited the villages on her own. Munk noticed that money was flowing into the villages, especially as measured by the number of people who built homes with metal roofs instead of thatch. But Munk and some of the people she interviewed for the book wondered if this relative prosperity would last after the project ended (Munk, 2013). I wondered too, so I looked for a more recent evaluation of the project and found one by Sachs himself, and his colleagues, published in 2018, based on surveys in 2015 at the end of the second and last 5-year phase of the Millennium Villages Project (Mitchell *et al.*, 2018).

The researchers saw some progress towards the UN's Millennium goals, especially for malaria, HIV/AIDS and maternal health. But the study found that the project had made no impact on poverty (Mitchell *et al.*, 2018).

It is a stunning admission, and I admire the team's honesty. Income in the Millennium Villages had increased a bit, but over the same decade most African economies had slowly improved. By the end of the project, the families in the Millennium Villages were no better off than households in the surrounding communities.

Paradoxically, the study found that the project had had a positive influence on agriculture, defined narrowly as the use of hybrid maize seed and chemical fertilizer, which Sachs and his team had encouraged, subsidizing it and distributing it to the local people (Mitchell *et al.*, 2018). And there's the rub.

The use of hybrid maize seed and chemical fertilizer may explain why the project did not end poverty. Expensive seed and fertilizer make farmers dependent on buying these inputs every year. If the rains fail one year, farmers may lose their maize, but those who bought seed and chemicals may also go into debt for these purchases. So, what Sachs's team thought of as a positive influence may have in fact undermined the potential of agriculture to contribute to poverty reduction.

Agriculture is also too complicated to reduce to simplistic, naïve solutions like hybrid seed and chemicals. Maize is a major crop in much of Africa, but not everywhere. Roots, tubers and bananas are more important for food security across much of the continent, especially the humid zones (Thiele *et al.*, 2017). As Munk (2013) describes for the village of Ruhiira, in south-west Uganda, although farmers did plant the maize seed and harvest it, they were unfamiliar with the crop. The locals didn't like to eat maize, had nowhere to store it and were not connected to grain buyers, making the grain difficult to sell.

Although Sachs was naïve and reductionist about agricultural development, I suspect that he was right about the need for governments and bilateral agencies to make massive investments in health, education and electricity. Governments have spent trillions of dollars to mitigate the pandemic lockdown, but for agriculture to help reduce poverty, mere investment is not enough. How the money is invested also matters. As explained in the report *Money Flows*, investments in agroecology are needed to contribute to build more resilient domestic food systems that could reduce the negative impact of short-term and long-term risks, and at the same time reduce poverty (Biovision Foundation for Ecological Development and IPES-Food, 2020).*

◼◀ Related video

Farmers in Africa and elsewhere can generate capital to set up microenterprises by organizing village savings and loan associations, as in this video from Malawi.

Udedi, R. (2019) *Village Savings and Loan Associations*. Hosted by Access Agriculture. 14 minutes.
www.accessagriculture.org/village-savings-and-loan-associations

---

*A version of the section above was previously published on www.agroinsight.com in 2020, by Jeff Bentley.

Of Fertilizers and Immigration

Chemical or mineral fertilizers have long been touted by agro-industry and by governments as a necessity to feed the growing world population. Sixty years after the start of the Green Revolution, the damage caused by industrial agriculture to farmland, surface water and groundwater, biodiversity and farmers' livelihoods has triggered many alarm bells. The application of pesticides and chemical fertilizers led to an increase in the level of heavy metals, especially cadmium, lead and arsenic, in the soil and groundwater, increasing the risk of cancer among humans (John and Babu, 2021; Sheng *et al.*, 2021). Environmental degradation and human health risks have led policy makers in India and in the European Union to curb the use of agrochemicals and encourage more environmentally friendly ways of farming (Khurana *et al.*, 2022). But fertilizers have also affected immigration in various ways.

Immigration can be triggered by political suppression or economic hardship, often aggravated by climate change. But rural folks across the globe are also under increased pressure due to the rising costs of agricultural inputs, such as chemical fertilizers and animal feed. While recently some European farmers have decided to migrate to other countries, the high rate of suicides among farmers in both Europe and India is shocking, with social issues, health problems, debts and crop failure cited as the major causes (Bossard *et al.*, 2016; Mariappan and Zhou, 2019; Bokinala, 2022). Despite these alarming events, the promotion of fertilizers in Africa goes on. As with the dumping of obsolete pesticides banned in Europe because of their high toxicity, the agro-industry, backed by institutions of the corporate food regime (Holt-Giménez and Altieri, 2012), has also turned to Africa to further increase their profits from selling fertilizers.

One of the problems is that for far too long, research has focused on yields instead of on farmers' profits, livelihoods and well-being, and on building healthy soils that can sustain farming in the long run. At a recent virtual conference organized by the European Commission, researchers from the Swiss Research Institute on Organic Agriculture (FiBL – *Forschungsinstitut für biologischen Landbau*) presented results from a 12-year study looking at various cropping systems in tropical countries. Soil organic carbon was on average 20–50% higher on organic farms than on conventional ones. While organic systems can yield as much or more than conventional systems, proper use of nitrogen-fixing legumes, organic manure and good agricultural practices are key to improving productivity (Bhullar *et al.*, 2021).

Fertilizer promotion by governments or development projects has mostly benefited local elites and better-off farmers, adding to social inequality. Modern cereal varieties have been bred for responsiveness to chemical fertilizer. At the beginning of the Green Revolution in the 1960s, rice, maize and wheat farmers who opted for the full package (modern high-yielding crop varieties, fertilizer and pesticides) initially were able to boost their yield. But while the increased production led to lower market prices, they also became increasingly indebted to moneylenders and banks (Kumar, 2020).

International researchers have now turned their attention to roots and tubers. The poor person's crop, cassava, could yield up to 35 tonnes of dry matter or nearly 100 tonnes of fresh roots per hectare, about four to five times the current average yield, if chemical fertilizers were used (Adiele *et al.*, 2020). Again, it will be mainly the larger farmers who stand to benefit as they capture the market. Smallholders stand to lose and, along with their children, turn to seek other livelihood options.

Cities in Africa are bursting with job seekers from the countryside, but migration outside the region offers even greater opportunities. According to the latest report of the International Organization for Migration (IOM, 2020, p. 318), land degradation, land tenure insecurity and lack of rainfall are major drivers of environment-induced migration for people from West and North Africa. The European narrative framing migration as primarily 'economic' often overlooks key factors, such as climate and environmental drivers of migration.

Environmental damage does not only happen where chemical or mineral fertilizers are used. It also happens where fertilizers are produced, but this remains often hidden.

Nauru, a Pacific island, was a good place to live when it gained independence from Australia in 1968. However, in just three decades of surface-mining, the island was stripped of its soil to get at the rock phosphate (for fertilizer). Now there is no place to grow crops. Ironically, Nauru's entire population has become dependent on imported food from Australia. More than 70% of Nauruans are obese, and the country struggles to reinstall backyard gardening and encourage young people to eat plant-based food. The mining of fertilizer and bad governance turned the smallest and once the richest republic in the world into the most environmentally ravaged nation on Earth: Nauru had little choice but to accept Australia's offer to host ousted asylum seekers, often immigrants from Indonesia, in return for money (LoFaso, 2014).

While some people and donors are still convinced that a Green Revolution industrial model of agriculture is the way forward for Africa, one should pause and look at the consequences of mining and using mineral fertilizer. If we want to create more opportunities for young rural people to earn a decent living, we have to support healthy food systems that nurture the soil and keep it healthy and productive.*

Related video

Organic biofertilizer is a viable, healthy alternative to chemical fertilizers.

Pagar, A. and WOTR (2020) *Organic Biofertilizer in Liquid and Solid Form*. Hosted by Access Agriculture. 15 minutes.
www.accessagriculture.org/organic-biofertilizer-liquid-and-solid-form

---

A version of the section above was previously published on www.agroinsight.com in 2021, by Paul Van Mele.

Fighting Farmers

Farmers belong to one of the world's most entrepreneurial professions. They have to deal not only with pests, crop diseases and the vagaries of climate, but also with fluctuations in market price and changing demands of retailers and preferences of consumers. Now a new threat is lurking on the horizon: farm machinery makers want to restrict the ability of farmers to mend their own machines, increasing costs and eating into farmers' already narrow profit margins.

Generations of farmers have tinkered with tools and machines to make work on the farm easier. Those days may become history soon. Under the Digital Millennium Copyright Act, a United States copyright law, manufacturers such as John Deere want to legally stop farmers across the globe from fixing their own machinery if the design of that machine involves electronic devices protected by copyright. An extract from a recent Farm Hack blog post, 'Farmers fight for the right to repair their own tractors', summarizes common fears about such property laws:

> While high-tech agricultural machinery has made the job of farmers more comfortable and more efficient in many regards, this same equipment has also proven to be a nightmare for farmers accustomed to equipment with simple control panels that don't resemble something found on the flight deck of the Starship Enterprise. A generation of farmers capable of popping open the hood and fixing a broken engine with their eyes closed now have their hands tied. While much of the gruelling work involved with farming has eased, so has a sense of control.

Complex, digitalized machinery designs and proprietary rights are hampering farmers' creativity and independence, but a community of fighting farmers has stood up. For instance, Farm Hack (farmhack.org) is an online community of farmers, designers, developers and engineers helping farmers to be better inventors. They develop and freely share tools that fit the scale and ethics of sustainable family farms. Another initiative, the crowdsourced magazine *Farm Show* (farmshow.com), showcases thousands of local farming inventions from the past three decades.

Initiatives such as Fairtrade, farm shops and other examples of short food supply chains offer a better and more reliable income to farmers, instilling a sense of connection with consumers while retaining the independence that farmers cherish. The ability to develop and share innovations in farm machinery is an equally important part of that independence and identity that sustains the passion of one of the oldest and most noble professions in the world.*

📹 Related video

Modern, appropriate farm machinery can be part of a profitable family farm.

Shanmuga Priya J. (2024) *A Hand Weeder and a Ridge Plough*. Hosted by Access Agriculture. 15 minutes.
www.accessagriculture.org/hand-weeder-and-ridge-plough

A version of the section above was previously published on www.agroinsight.com in 2016, by Paul Van Mele.

Family Farms Produce More Food and Jobs

Family farms grow more food and employ more people than industrialized farms, which do a poor job at both. In Bolivia, big commercial farms cut down prime forest while avoiding taxes, as shown in a recent study (Czaplicki Cabezas, 2021) conducted for an old and respected NGO, Centre for Research and Promotion of the Peasantry (CIPCA – *Centro de Investigación y Promoción del Campesinado*).

The study, based on the 2015 agricultural census and on later data, classifies farms into the following four categories:

1.   Non-family. Produce mainly for export, using machinery and hired labour. Main crops are soy beans, sunflower, rice and sugar.
2.   Consolidated family farms. Use hired and household labour to produce coffee, coca leaves, bananas and plantains, maize, wheat, soy beans and other crops for domestic and export markets.
3.   Transitional family farms. Produce fruit, maize, wheat and other crops, mostly with household labour, usually for the local market.
4.   Subsistence farms. Families use their own labour to produce potatoes, other roots and tubers, broad beans and peas.

Main jobs

Agriculture provides 1.45 million paid jobs, or 'contracts', in Bolivia (including part-time and seasonal jobs). Another 1.77 million people work on their own farms. Non-family farms create only 125,720 jobs, less than 9% of the total. Non-family agriculture needs 6.5 ha to create a paid job, while family farmers (mainly the consolidated ones) create one paid job for every 1.6 ha, mainly because they use less machinery.

Crops

Soy beans have become Bolivia's main crop by land area, and they are now produced on 37% of the country's farmland, much of it recently cleared from the forest. Gravetal, a large company that exports soy, has only 195 employees, and figures are similar for other large companies. When soy prices peaked (2012–2014), the six largest soy processing and export companies paid less than 1% of the value of their sales in taxes. Almost all of the soy is used to feed livestock, mostly outside of Bolivia. At the other extreme, family farms produce 99% of Bolivia's potatoes, the main food crop. In other words, large farms hardly generate revenues for the country and they do not feed the country.

I've summarized the study in Table 3.1, to show that family farms produce more food per hectare than do non-family farms. 'Food' does not include the soy beans fed to animals. Subsistence farms produce the most food per hectare. The industrialized, non-family farms produce just 12.4% of the food on 38% of the land, a serious inefficiency in the food system. Family farmers produce four times as much food per unit of land as do industrialized farms.

This is an important quantitative study, showing that family farms produce more food and employ more people than does industrial agriculture. Small farms make more efficient use of land than industrial agriculture. Family farms deserve more investment, research, technical advice and other support.*

Table 3.1. Land and food production in Bolivia on non-family and family farms. Adapted from Czaplicki Cabezas (2021).

	Farmland (%)	Farmland (1000 ha)	Food produced (%)	Food production (tonnes)	Tonnes of food per hectare
Family subsistence	12	432	23.7	572,055	1.32
Family, transition	19	684	29.5	712,639	1.04
Family, consolidated	31	1116	34.5	835,260	0.75
Non-family	38	1368	12.4	298,714	0.22
Total		3600		2,418,668	0.67

▶ Related video

In this video from Bolivia, smallholders explain how they conserve natural vegetation around farms to provide feed and habitat for a wasp, which is a natural enemy of a serious crop pest. Practices like this help to reduce pesticide abuse on small farms.

Agro-Insight and PROINPA (2019) *The Wasp that Protects Our Crops*. Hosted by Access Agriculture. 9 minutes.
www.accessagriculture.org/wasp-protects-our-crops

A version of the section above was previously published on www.agroinsight.com in 2021, by Jeff Bentley.

Repurposing Farm Machinery

Many farmers in Europe and North America are burdened with debts due to the heavy investments they have made over the years to buy farm machinery.

A new tractor easily costs 100,000 Euros or more. New agricultural policies often force farmers to change as well. When environmental policy outlawed the spreading of liquid manure on the surface of the field, manufacturers quickly adapted: manure is now directly injected into the soil. But this may oblige farmers to get rid of machinery that still works. What solutions can research offer to repurpose farm equipment? These thoughts have gradually come to my mind, living in a farming village in north-eastern Belgium and observing the various changes.

Farmers creatively adapt in many ways. Our friend Johan Hons uses a leek planter to transplant sweet maize seedlings on his organic farm to reduce the need for weeding. Like many farmers, Johan has his own workshop where he adjusts equipment to suit his needs.

American and European farmers see the soaring prices of equipment as one of the key challenges adding to their debt. Besides, equipment has become so complicated and repair is stymied by proprietary software and a lack of available parts. As a response, many farmers are now buying simpler and much cheaper second-hand tractors from the 1970s and 1980s.

Also, local service providers have repositioned themselves and taken over many of the farm operations (Fig. 3.1). And the fewer local service providers there are, the more pressure they can put on farmers, often charging fees that further eat into farmers' meagre profit margins. Many machines, like the ones that inject liquid manure into the soil, have become so big that they are wider than the country lanes, damaging them and forcing cyclists to jump off the road to save their lives when these machines roar by.

Due to tillage and the use of agrochemicals, many soils have become depleted of organic matter and soil life. As agricultural policies for decades have supported industrial agriculture, all farmers own their own pesticide spraying equipment (Fig. 3.2). So, will these become obsolete when farming transitions to more sustainable models? Or could pesticide spraying machines be used to spray the soils and crops with Effective Microorganisms® or other natural biofertilizers, to bring life back into our soils and boost crop health in a natural way?

To enable the transition to more sustainable farming, appropriate machines will be required. In the Netherlands, Wageningen University & Research has been studying intercropping for several years, involving conventional and organic farmers. By growing a variety of crops in narrow strips they were able to attract beneficial insects and slow the spread of crop disease. The researchers also found that yields are similar to those found in monocultures and labour requirements are comparable, too (van Apeldoorn, 2020). Reading their study, I immediately thought how intercropping would work in a highly mechanized setting. Adjusting machinery will likely be part of the solution.

With the action plan laid out in the European Green Deal, the EU aims to be climate neutral by 2050 (European Commission, 2019). Different sectors of society each have a responsibility to make this happen. For agriculture, the Farm to Fork Strategy stipulates that by 2030 pesticide use has to be reduced by 50% and chemical fertilizers by 20% in order to make food systems more sustainable.

Clearly, equipment manufacturers will continue to adjust the design of machinery, but this also comes at a cost. To keep as many farmers in business as possible, some creative thinking will be required on how to strike a balance between supporting industry to innovate and finding ways to repurpose the existing machinery that farmers have already invested in. European family farmers are ready to adapt, but they are also being run out of business. Policy and research should lend them a hand, by inventing and promoting appropriate small machinery that can be used to serve multiple purposes. In addition, policies will need to be put in place to protect farmers from data capture by large corporations, as farm machinery becomes increasingly digitalized.*

◼️ Related video

Somewhere between the backbreak of hand weeding and monster machines is the alternative of appropriate machinery. This video from West Africa is one of several on such implements.

AfricaRice and Real2Reel (2016) *Rotary Weeder*. Hosted by Access Agriculture. 18 minutes. www.accessagriculture.org/rotary-weeder

---

A version of the section above was previously published on www.agroinsight.com in 2020, by Paul Van Mele.

From Uniformity to Diversity

Industrial agriculture has so damaged our farmland that the survival of future generations is at risk, reveals Professor Emile Frison in his report *From Uniformity to Diversity*, but there is a way forward (IPES-Food, 2016).

Frison's conclusions are staggering. The indiscriminate use of synthetic fertilizers has destroyed the soil biota and its nutrient-recycling potential. The combination of monocultures with highly mechanized farming and fertilizer abuse has caused historical land degradation on over 20% of the Earth's agricultural land (IPES-Food, 2016).

High-yielding varieties and abundant chemical inputs increased global crop yields in the early decades of the Green Revolution, but by now the sobering figures indicate that productivity has stagnated or collapsed in 24–39% of the areas that grow maize, rice, wheat and soy beans (Ray *et al.*, 2012).

The productivity of industrial agriculture has systematically degraded the environment on which it relies. The use of pesticides in agriculture has caused a global decline in insect pollinators, threatening the very basis of

Fig. 3.1. In Western Europe, service providers with specialized, computerized equipment have taken over many of the field activities.

agriculture. Some 35% of global cultivated crops depend entirely on pollination by insects (IPES-Food, 2016).

Pests, diseases and weeds are adapting to chemical pest management faster than ever. Genetically modified soy bean and maize that are herbicide-tolerant have led to an indiscriminate use of glyphosate-based herbicides such as Roundup® and 2,4-D. Some 210 species of weeds have now evolved resistance to herbicides. Clearly, this flawed industrial model has mainly benefited corporate interests and the wealthiest farmers.

Of equally great concern to our future generations, industrial agriculture significantly reduces the agrobiodiversity of livestock and crops. Underutilized or minor crops such as indigenous leafy vegetables, small-grained African cereals, legumes, wild fruits and tree crops are disappearing in the face of competition with a few industrially produced varieties of rice, maize and wheat (IPES-Food, 2016).

Greenhouse gases, water pollution, over-exploited aquifers, soil erosion, loss of agrobiodiversity and epidemics such as avian influenza and foot-and-mouth disease are all signs that we need to urgently rethink the way we produce, source and consume food.

Fig. 3.2. Without stringent policies and control measures, pesticide spraying will continue to damage the environment and human health.

A study covering 55 crops grown on five continents over 40 years found that organic agriculture was 22–35% more profitable than conventional agriculture (Reganold and Wachter, 2016).

In developed countries, yields of organic agriculture were 8% lower than conventional agriculture, but they were 80% higher in developing countries where the negative impacts of industrial agriculture on food and nutrition security are felt much more strongly. Data from temperate and tropical agroecosystems suggest that legume cover crops could fix enough nitrogen to replace the amount of synthetic fertilizer currently in use (Badgley *et al.*, 2007).

So, diversified systems have shown the capacity to raise productivity in places where additional food is desperately needed. Yet corporate lobby groups, some donors and development agencies continue to push governments towards unsustainable production models. In many developing countries, the general switch towards specialized, export-oriented systems has eroded the diverse farming economy, causing a gradual loss of local food distribution systems (Gliessman, 2007). With rapid shifts in global and regional competitiveness this has destabilized national food supply, not only jeopardizing the very livelihoods on which rural people depend, but also putting the economic and political stability of developing countries at risk.

Ethical labels, such as Fairtrade, ensure that farmers in developing countries get more money for their produce, while at the same time ensuring that social and environmental services are ploughed back into the rural communities, as explained by Nicolas Lambert, CEO of Fairtrade Belgium (personal communication, Nicolas Lambert, Belgium).

Emile Frison and other outstanding scientists like Professor Olivier De Schutter, former United Nations Special Rapporteur on the right to food, have joined forces in the International Panel of Experts on Sustainable Food Systems (IPES-Food). There is indeed an urgent need to alert policy makers to the high risks related to short-term thinking and concentration of power in the hands of fewer, large-scale retailers and corporate agri-businesses.

It is reassuring that eminent people have teamed up to protect global biodiversity and farmers' rights to seed as key requirements for food systems that respect the farmers and their environment. The opponents are powerful, and motivated by greed, so the struggle is bound to be a long one.*

◼️ Related video

Farmers can be helped to organize community seed banks, to defend their rights to plant their own seed.

Agro-Insight (2023) *Community Seed Banks*. Hosted by Access Agriculture. 15 minutes. www.accessagriculture.org/community-seed-banks

---

A version of the section above was previously published on www.agroinsight.com in 2018, by Paul Van Mele.

Making Farmers Anonymous

Short food chains narrow the gap between producers and consumers, but when food is traded through wholesalers and when supermarkets sell their own brands, active efforts are undertaken to make farmers anonymous. This never really occurred to me until Vera Kuijpers, from the organic farm and farm shop Het Eikelenhof in north-eastern Limburg, told me about her latest experience with Biofresh, one of the major organic wholesalers in Belgium.

Every Thursday at 4:30 AM, Vera and her husband, Johan Hons, load their van with their freshly harvested vegetables and other produce and drive about 100 km to sell to Biofresh, among a few other places. One day in 2021, they also took over 60 crates of new potatoes. Their red Aloette is a firm and delicious potato variety that withstands the common Phytophthora disease. Each crate is nicely labelled with the family name of Johan (Hons) and the variety. 'The moment we deliver our crates, staff at Biofresh remove the labels and attach their own, so that other people who come and buy organic produce will not know who has produced it. But this time, they also removed the name of the variety,' Vera says, 'they just put "red potato". I really wonder why they would do that.'

Vera and Johan strongly disagree with this new development, as one cannot compare one red variety with all the others. 'If customers in a supermarket want to buy the delicious Aloette, one week it may be this variety, but if the next week it is another red variety that has a very different taste or is not tasty, they may stop buying red potatoes altogether, and then the farmers who grow Aloette will no longer be able to sell their potatoes,' Johan says. From a producer's perspective he surely has a point.

When discussing this matter with my wife, Marcella, she points out the stark contrast with how Oxfam markets Fairtrade products. On almost all packaging, you see the face of a farmer, often with a personal story. Supermarkets also sell Fairtrade products and have no problem with an occasional personal touch. A coffee drinker in Europe is unlikely to meet many coffee growers in the distant African highlands or Latin America. The photo and the blurb on the coffee package are a mere virtual connection. There is little risk that European consumers will buy directly from growers in Kenya or Colombia, but a Belgian retailer or consumer would be able to contact a farmer in Belgium.

Naming the farmers and the varieties is important when selling produce, as it helps consumers relate to the food they eat and the people who produce it. As consumers increasingly turn to farm shops, farmers' markets and other short food chains, wholesalers and supermarkets are now taking steps to make farmers and varieties anonymous. The farmers are potential competitors, and the big buyers want to make sure that the customers never meet the farmers. It is like a trade secret to stay in business, as well as yet another tactic whereby farmers' ability to negotiate prices with middlepersons and supermarkets is undermined.*

▣ Related video

During the COVID-19 lockdown, farmers in India organized to deliver organic produce, which consumers ordered online.

Pagar, A. (2022) *Home Delivery of Organic Produce*. Hosted by Access Agriculture. 15 minutes.
www.accessagriculture.org/home-delivery-organic-produce

A version of the section above was previously published on www.agroinsight.com in 2021, by Paul Van Mele.

References

Adiele, J.G., Schut, A.G.T., Van Den Beuken, R.P.M., Ezui, K.S., Pieter Pypers, A.O. *et al.* (2020) Towards closing cassava yield gap in West Africa: Agronomic efficiency and storage root yield responses to NPK fertilizers. *Field Crops Research* 253, 107820.
Badgley, C., Moghtader, J., Quintero, E., Zakem, E., Chappell, M.J. *et al.* (2007) Organic agriculture and the global food supply. *Renewable Agriculture and Food Systems* 22, 86–108.
Belgische Boerenbond (1990) *100 Jaar Boerenbond in Beeld. 1890–1990 (An Overview of 100 Years of Boerenbond in Pictures. 1890–1990)*. Dir. Eco-BB – S.Minten, Leuven, Belgium.

Bhullar, G.S., Bautze, D., Adamtey, N., Armengot, L., Cicek, H. *et al.* (2021) *What Is the Contribution of Organic Agriculture to Sustainable Development? A Synthesis of Twelve Years (2007–2019) of the 'Long-Term Farming Systems Comparisons in the Tropics (SysCom)'*. Research Institute of Organic Agriculture (FiBL), Frick, Switzerland. Available at: https://orgprints.org/id/eprint/39536/1/Syscom_Synthesis_Report.pdf (accessed 16 October 2024).

Biovision Foundation for Ecological Development and IPES-Food (2020) *Money Flows: What Is Holding Back Investment in Agroecological Research for Africa?* Biovision Foundation for Ecological Development & International Panel of Experts on Sustainable Food Systems. Available at: https://www.ipes-food.org/_img/upload/files/Money%20Flows_Full%20report.pdf (accessed 16 October 2024).

Bokinala, V.R. (2022) Investigation into the causes and potential mitigation of the high incidence of farmers' suicide. Doctoral dissertation, University of Surrey, UK.

Bossard, C., Santin, G. and Canu, I.G. (2016) Suicide among farmers in France: Occupational factors and recent trends. *Journal of Agromedicine* 21, 310–315.

Czaplicki Cabezas, S.T. (2021) *Desmitificando la Agricultura Familiar en la Economía Rural Boliviana: Caracterización, Contribución e Implicaciones (Demystifying Family Agriculture in the Bolivian Rural Economy: Characterization, Contribution and Implications)*. Centro de Investigación y Promoción del Campesinado (CIPCA), La Paz, Bolivia.

European Commission (2019) The European Green Deal. Available at: https://ec.europa.eu/info/strategy/priorities-2019-2024/european-green-deal_en (accessed 16 October 2024).

Eurostat (2022b) EU Agricultural Prices Continued to Rise in Q3 2022. Available at: https://ec.europa.eu/eurostat/web/products-eurostat-news/w/ddn-20221222-2 (accessed 16 October 2024).

Gliessman, S.R. (2007) *Agroecology: The Ecology of Sustainable Food Systems*. CRC Press, Boca Raton, Florida.

Holt-Giménez, E. and Altieri, M.A. (2012) Agroecology, food sovereignty, and the new Green Revolution. *Agroecology and Sustainable Food Systems* 37, 90–102.

IOM (2020) *Migration in West and North Africa and across the Mediterranean: Trends, Risks, Development and Governance*. International Organization for Migration (IOM), Geneva, Switzerland.

IPES-Food (2016) *From Uniformity to Diversity: A Paradigm Shift from Industrial Agriculture to Diversified Agroecological Systems*. International Panel of Experts on Sustainable Food systems. Available at: https://ipes-food.org/report/from-uniformity-to-diversity/ (accessed 16 October 2024).

IPES-Food (2019) Towards a Common Food Policy for the EU. Available at: www.ipes-food.org/pages/CommonFoodPolicy (accessed 16 October 2024).

John, D.A. and Babu, G.R. (2021) Lessons from the aftermaths of Green Revolution on food system and health. *Frontiers in Sustainable Food Systems* 5, 644559.

Khurana, A., Halim, M.A. and Singh, A.K. (2022) *Evidence (2004–20) on Holistic Benefits of Organic and Natural Farming in India*. Centre for Science and Environment, New Delhi, India.

Kumar, S. (2020) *Agrarian Crisis and Farmers Issues in India: From Indebtedness to Suicide*. Best Publishing House, New Delhi, India.

Livingstone, E. (2018) Europe's Hidden Milk Lake Threatens Fragile Market. Politico. Available at: www.politico.eu/article/europes-hidden-milk-price-lake-threatens-fragile-market-eu-commission/ (accessed 16 October 2024).

LoFaso, J. (2014) Destroyed by Fertilizer, a Tiny Island Tries to Replant. Modern Farmer. Available at: https://modernfarmer.com/2014/03/tiny-island-destroyed-fertilizer-tries-replant/ (accessed 16 October 2024).

Mariappan, K. and Zhou, D. (2019) Threat of farmers' suicide and the opportunity in organic farming for sustainable agricultural development in India. *Sustainability* 11, 2400.

Mitchell, S., Gelman, A., Ross, R., Chen, J., Bari, S. *et al.* (2018) The Millennium Villages Project: A retrospective, observational, endline evaluation. *Lancet Global Health* 6, e500–e513.

Munk, N. (2013) *The Idealist: Jeffrey Sachs and the Quest to End Poverty*. Anchor Books, New York.

Ray, D.K., Ramankutty, N., Mueller, N.D., West, P.C. and Foley, J.A. (2012) Recent patterns of crop yield growth and stagnation. *Nature Communications* 3, 1293.

Reganold, J.P. and Wachter, J.M. (2016) Organic agriculture in the twenty-first century. *Nature Plants* 2, 15221.

Sheng, D., Wen, X., Wu, J., Wu, M., Yu, H. *et al.* (2021) Comprehensive probabilistic health risk assessment for exposure to arsenic and cadmium in groundwater. *Environmental Management* 67, 779–792.

Statbel (2023) Land- en tuinbouwbedrijven (Farm and horticultural holdings). Available at: https://statbel.fgov.be/nl/themas/landbouw-visserij/land-en-tuinbouwbedrijven#figures (accessed 16 October 2024).

Steck, N., Junker, C., Bopp, M., Egger, M. and Zwahlen, M. (2020) Time trend of suicide in Swiss male farmers and comparison with other men: A cohort study. *Swiss Medical Weekly* 150, w20251.

Thiele, G., Khan, A., Heider, B., Kroschel, J., Harahagazwe, D. *et al.* (2017) Roots, tubers and bananas: Planning and research for climate resilience. *Open Agriculture* 2, 350–361.

van Apeldoorn, D. (2020) More nature in fields through strip cropping. Wageningen University and Research Spotlight, 22 May. Available at: https://weblog.wur.eu/spotlight/more-nature-in-fields-through-strip-cropping/ (accessed 16 October 2024).

PART 2

Diverse and Ecological Farming as a Viable Alternative

Family farmers are stewards of the land. For example, early Mesopotamians farmed sustainably for 5000 years before states developed (Scott, 2017). On the other hand, large corporations 'invest' in agriculture to strip the land of its value before moving on (Philpott, 2020). Family farmers often raise diverse crops, trees and animals, and by doing so conserve biodiversity. Smallholders who produce their own seed also keep their costs down, which is important, as we have seen in Part 1. There are already many techniques for conserving farmland and water, and for managing crop pests and diseases in ecologically sound ways. Animals and trees are key elements of an integrated, profitable small farm.

References

Philpott, T. (2020) *Perilous Bounty: The Looming Collapse of American Farming and How We Can Prevent It*. Bloomsbury Publishing, New York.

Scott, J.C. (2017) *Against the Grain: A Deep History of the Earliest States*. Yale University Press, New Haven, Connecticut.

4   Biodiversity

Abstract

Agroecology fosters biodiversity and sustainability in farming practices worldwide: Francisco 'Pacho' Gangotena's organic, mixed farm in Ecuador achieves high yields while sequestering carbon, protecting birds and selling direct to consumers. The Ortega family on the Bolivian Altiplano integrates traditional practices with innovations like biofertilizers and greenhouses. Farmers in Cochabamba, Bolivia, implement agroforestry, planting trees alongside crops to improve soil health and create a diverse ecosystem. Development organizations in West Africa promote integrated farming systems that combine tree crops (cashew, mango) with legumes (groundnuts, soy) to provide food security during the lean season and discourage bush fires. Belgian organic farmers Johan Hons and Vera Kuijpers cultivate forgotten vegetables like rhubarb, preserving biodiversity and creating a unique market niche.

Enlightened Agroecology

Francisco 'Pacho' Gangotena grew up in the countryside of Ecuador and decided that the best way to help smallholder farmers was to get an education. So, he went abroad for a PhD in anthropology. He came home feeling like 'the divine papaya', he says, thinking that he could change the world with his doctorate.

After a year of teaching at the university, Pacho wanted to do something more practical, so he and his wife, Maritza, sold the house and the car and bought 4 ha of land for farming not too far from Quito. But making this work was going to be a huge challenge. The land had no trees and the soil was degraded.

From day one, the family decided that they would use no agrochemicals. They gradually improved the soil by recycling the crop residues and manure back into the soil. Pacho estimates that in this way the family has applied the equivalent of 4000 truckloads of compost since he first began farming here over 35 years ago.

I met Pacho recently on his farm in Puembo, in the Ecuadorian Andes, where he happily showed me and a few other visitors his four dairy cows. He puts sawdust in their stall to absorb their manure and urine. Every day, each cow eats 90 kg of feed and produces about 70 kg of waste, which amounts to about 25 t of organic fertilizer each year from each cow. A single cow can fertilize 1 ha of crops. All the manure goes onto the farm, along with all of the composted crop residues.

Pacho rotates his vegetable crops on his 4-ha farm. Potatoes are followed by broccoli, lettuce, radishes and green beans. He employs ten people and is proud that his small farm can give jobs to local families by producing healthy vegetables to sell directly to consumers in the local markets.

His grown son and daughter also work on the farm. Pacho jokes that he has retired and that now his daughter is his boss – and a pretty demanding one.

Besides recycling organic matter, Pacho also has some more unusual strategies for building up the soil. He enriches it with wood ash from pizzerias and with powdered rock from quarries. As the quarries cut stone, they leave behind a lot of powdered rock as waste, which Pacho collects. Rocks are rich in minerals (with up to 80 elements) and are one of nature's main components of soil.

Pacho is up front about his limitations, which adds to his credibility. A new phytoplasma disease (*punta morada*) is sweeping across Ecuador, wiping out potato fields, including his. He also has to import vegetable seed from the USA and Europe.

But Pacho's vegetable fields are lush, like gardens, and now surrounded by trees that the family has planted 'providing room, board and employment for the birds and for beneficial insects', Pacho explains. An ornithologist friend counted 32 bird species on the farm, including 22 insectivores. Pacho is convinced that the birds help him to control pests without insecticides. Predatory insects also control pests.

He also thinks that it is important to share what he has learned, and has welcomed around 32,000 smallholders to visit his farm over the years. It helps that he was the director of SWISSAID in Ecuador for 20 years and has built a large network of collaborating farmers. Many come in groups, and some stay for several days to learn about organic farming and agroecology.

The farm family and staff feed us a big lunch of kale salad, potato soup and a lasagna made with green leaves instead of pasta. All vegetarian and delicious. The farm has a clear emphasis on nutritious food and produces lots of it. By intercropping and rotating crops, they get 92 t of vegetables and other crops per hectare each year, a more than respectable yield. This is four times the average world vegetable yield of just 18.9 t per hectare (Dong *et al.*, 2022). Since buying the farm, the organic matter, or carbon held in the soil, has increased from 2 to 12% or more. To put this in perspective, 6% soil carbon is considered high in North American farms (Montgomery, 2017). A 1% increase of soil carbon is about 21 t per hectare (Toensmeier, 2016, p. 22).

Not everyone is in favour of organic agriculture. For example, in an otherwise excellent book, *Enlightenment Now*, Pinker (2018) argues that organic agriculture is not sustainable because it supposedly uses more land than conventional agriculture. In fact, in developing countries organic agriculture yields 80% more than conventional agriculture (IPES-Food, 2016).

That brings us back to the Gangotena family farm, which is providing jobs, and lots of healthy food, while removing carbon from the air where it is harmful and putting it underground where it is useful. Organic and other forms of ecological agriculture, such as regenerative agriculture and agroforestry, may be one of the world's feasible and cost-effective ways for sequestering carbon from the atmosphere, storing it in the soil as rich, black earth for productive farming. A farm can be profitable, and a good place to work, with a diversity of crops and trees that also supports many species of birds (and no doubt other animals as well).*

▶️ Related video

Smallholders and researchers in Peru collaborated to create a pasture system that provides more fodder and improves the soil.

Agro-Insight and Yanapai (2022) *Improved Pasture for Fertile Soil.* Hosted by Access Agriculture. 15 minutes.
www.accessagriculture.org/improved-pasture-fertile-soil

A version of the section above was previously published on www.agroinsight.com in 2019, by Jeff Bentley.

Harsh and Healthy

Hours away from any city and a half-hour drive from the paved highway, we meet don Miguel Ortega, a warm, welcoming man in his late 40s, along with his wife, Sabina Mamani, and three of their five children on their farm in Viloco village. In this remote area on the northern Altiplano of Bolivia, I wonder how he manages to feed his family. But first impressions can be deceiving. Later in the day we meet his daughter who studies at the university and I realize that this is a prosperous family that is investing in education and healthy food.

The landscape is quite unlike the southern Altiplano, where the sandy soils and the mere 150 mm of rainfall per year allow farmers to only grow quinoa and rear llamas and sheep. Here, further north, there are more options; soils are more fertile and, with 500 mm of rainfall, farmers grow quinoa, potatoes, broad beans, barley and lucerne (alfalfa) as fodder. Dairy cows are as prevalent as llamas.

Fig. 4.1. Don Miguel Ortega explaining the *Pachagrama*, a locally invented method to record weather and cropping information.

Don Miguel is one of the 70 *Yapuchiris*, experienced farmers on the Altiplano who share their skills with their peers (Quispe *et al.*, 2008, 2018). He is hired by several NGOs to train groups of farmers on ecological farming, including how to make organic inputs such as biol (fermented liquid manure), and how to fill out the *Pachagrama*, a locally invented method to record natural weather indicators and a cropping calendar so farmers can make better decisions (Fig. 4.1).

Don Miguel's home, a cluster of adobe buildings, houses animals and vegetables that produce a tasty and healthy diet. The farm also has three neo-Andean greenhouses, made with adobe walls and topped with yellow agro-film, a tough plastic that withstands the sun (Fig. 4.2). But one greenhouse is not used to grow vegetables. It turns out to be a homemade biogas installation. The greenhouse structure ensures that the manure and organic waste keeps fermenting during the cold winter months. The unit provides the family year-round gas to cook for 2 hours per day. Being off the grid, a solar panel supplies the household the minimum amount of electricity.

Mid-morning, one of the young girls brings us a mandarin orange. We accept the fruit with a sense of wonder. At nearly 4000 m above sea level there are hardly any trees, certainly none that require mild Mediterranean

Fig. 4.2. Growing vegetables in a greenhouse at an altitude of about 4000 m.

temperatures. When don Miguel invites us in one of his greenhouses, we see a single mandarin tree with a few fruits.

In the greenhouse he opens a black plastic sheet laying on the soil. Hundreds of earthworms seek shelter from the light, crawling deeper into the decomposing manure. He tells us that he watched a video a while ago from Bangladesh where farmers were also rearing earthworms. The video had been translated into Aymara and Spanish. While don Miguel had been rearing earthworms before he saw the video, he was pleasantly surprised to see farmers growing earthworms on the other side of the world, and he realized that in the future he could perhaps make enough vermicompost to have some to sell. Training videos from other countries not only give farmers new ideas, they also give them confidence in their own innovations and practices.

The family treated their visitors to a delicious, traditional Andean meal with mutton, potatoes and *chuño* (potatoes that are freeze-dried outside during the winter nights). Unusual for a household on the Altiplano, they also serve organic, leafy vegetables, fresh from the greenhouse. All comes with a delicious yellow sauce, which later on we are told is prepared by their teenage son who aspires to become a chef one day.

It is often stated that people in remote areas only grow organic crops by default, because they cannot afford chemical fertilizers and pesticides (Juroszek *et al.*, 2008). Don Miguel and the many *Yapuchiris* we have met confirm that such statements are an insult to the many farmers who decide to live in harmony with nature, with care for their environment, their health and their families. Farmers in remote areas can be enabled to learn from their peers, to produce a diversity of new crops, while keeping the local ones that they love.*

▶ Related video

In the following video don Miguel and other Bolivian farmers show how they record and follow their seasonal weather forecasts.

Agro-Insight and PROSUCO (2019) *Recording the Weather*. Hosted by Access Agriculture. 14 minutes.
www.accessagriculture.org/recording-weather

A version of the section above was previously published on www.agroinsight.com in 2018, by Paul Van Mele.

Farming with Trees

On a rocky hillside an hour from the city of Cochabamba, agronomist Germán Vargas points out a molle tree (*Schinus molle*). It's growing from a crack in a sandstone boulder with little or no soil. Native trees are well adapted to their environment and don't need much to survive, Germán observes.

Molle can be cut for good firewood, but it also casts an inviting shade, with a thick carpet of fallen leaves. Trees grown on farms also have multiple uses. Some have deep roots that bring up nutrients from beneath the top soil. Even in places like Cochabamba, with a long dry season, many trees stay green all year round. The trees have found water to keep their leaves moist, despite the bone-dry subsoil. Germán explains that farming with trees, or agroforestry, mimics natural forests, where rich soils are created without irrigation or fertilizer.

Four years ago, Germán and two colleagues bought some land to put their ideas on agroforestry into practice. They now have 1500 apple trees in a 4-ha orchard, on a former onion farm, where the intensive use of chemical fertilizers and pesticides had depleted the soil of nutrients.

Germán and his friends bought some apple seedlings from a local nursery. They chose improved Brazilian apple varieties, such as Eva and Princesa, which do well in the highland tropics of South America, where it can get cool but does not freeze. Germán learned about apple farming from Serafín Vidal, an agronomist who is also an extension agent.

Germán and his colleagues plant a few more trees every year. They start each new planting by digging a trench every two to three metres (depending on the slope), to let water infiltrate the soil. They throw the soil just uphill of the trench to create a barrier, slowing down the runoff of water and trapping sediment.

Germán is careful not to scrape the soil surface with hand tools; the top soil is so thin that rough handling could remove it all. They add a little compost to the soil, mimicking a natural forest, where fallen leaves and trees rot and release nutrients back into the soil. However, forests also have an understory, so potatoes, maize, lettuce, amaranth, rye and other plants are sown between the trees. After the vegetables have been planted, a straw mulch keeps down the weeds.

Other trees are planted among the apples, to make leaf litter, add nutrients and help keep the soil moist. Germán experiments with natives like molle and exotic species. For example, he brought seed of the *chachafruto* tree (*Erythrina edulis*) from Colombia. The plant is adapting well. When the only date palm in Cochabamba, another non-native species, dropped a cluster of dates in a city park, Germán salvaged the seed and planted some on the farm.

The apples were remarkably free of mildew, mites, fruit flies and other common pests, but even if they were to appear, Germán avoids using pesticides. The team makes a spray with cow manure, raw sugar, bone meal, sulfur, ash and lime. Reasoning that all stone has mineral nutrients, they add a little 'rock flour', made by grinding a soft local sedimentary stone. A culture of beneficial microorganisms is added to ferment the mix in sealed drums. The agroforesters culture the microorganisms themselves, but they get the starting culture in the local forest, bringing in a few handfuls of soil with semi-decomposed fallen leaves. The sulfur and the lime come from the farm supply store. This sulfur blend is sprayed about five times a year on the trees, to fertilize them and control pests. It seems to be working, since the apples have almost no pests, except for birds, and the annual plants are thriving.

This innovative agroforestry system needs regular attention and it is obviously a lot of work, especially at first, because it is established by hand, without machinery. Some of the radishes have gone to seed, and in a few beds the weeds are lush and healthy, waiting to be cut down for the next vegetable crop.

Agroforestry has added biodiversity to what was once a barren farm. Farmers can learn from forests to make better use of water, conserve the soil and manage pests and disease naturally, thanks to a diversity of plants that includes trees. Farming with trees can yield a good harvest of fruits and vegetables, while building and sustaining soils (Toensmeier, 2016).*

◾️📹 Related video

Bolivian expert Ing. Serafín Vidal explains agroforestry in this video.

Agro-Insight and PROINPA (2023) *Seeing the Life in the Soil*. Hosted by Access Agriculture. 9 minutes.
www.accessagriculture.org/seeing-life-soil

A version of the section above was previously published on www.agroinsight.com in 2020, by Jeff Bentley.

A Burning Hunger

Towards the end of the dry season, many families across the African savannas have exhausted their reserves of stored cereal crops. Vegetables are hard to come by in local markets. Bush meat is one way for rural people to supplement their meagre diet with protein during the well-named lean or hunger season. This is why development organizations have struggled for decades to curb the destructive practice of setting the bush on fire to hunt small wildlife (Fig. 4.3).

One option to ensure some food and income during the lean season is to grow cashew and mango trees. But with increased labour costs and insecure markets, farmers find it hard to properly maintain their planted trees. Slashing the weedy and bushy undergrowth is often only done late during the flowering and fruiting season, by which time bush fires set by others may have spread into and destroyed entire groves in no time.

Increasingly, development organizations are starting to realize that integrated farming systems and local value addition to food are the way

Fig. 4.3. Bush fires in West Africa are traditionally used to hunt for bush meat.

forward. In a published video on the Access Agriculture video platform, the Beninese NGO the Organization for Sustainable Development, Strengthening and Self-Promotion of Community Structures (DEDRAS – *Organisation pour le Développement Durable, le Renforcement et l'Autopromotion des Structures Communautaires*) neatly shows how growing groundnuts and soy beans in cashew plantations helps farmers produce a nutritious crop during the lean season, and thus discourages damaging bush fires. DEDRAS also made a training video with rural women on how to make cheese from soy, a good example of adding value (Fig. 4.4).

In addition to tree crops, such as mango and cashew, farmers also manage other local species, such as néré (*Parkia biglobosa*) and the karité or shea nut tree (*Vitellaria paradoxa*). These wild indigenous trees, distinctive features of the savanna, also provide fruits and nuts during the lean season. Néré and the shea nut tree have grown here for thousands of years and are relatively fire-resistant. Traditionally, néré seeds are dried, cooked and fermented to make 'soumbala', a local equivalent to bouillon cubes that brings taste to many dishes. But with an increased need for fuel wood, more néré are being cut down. While the fuel wood crisis has not received the attention it deserves, nutritionists have taken notice and have come up with a way to use fermented soy beans as a replacement for the local soumbala. This practice

Fig. 4.4. Alternative protein sources reduce the need for bush fires.

has been captured by the NGO AMEDD in Mali in a nice farmer training video, also hosted on the Access Agriculture video platform.

In West Africa, declining populations in the wild, along with the strong and continuing demand for meat, have inspired rural entrepreneurs to develop alternative sources. Across Africa one can witness how mainly women and youth have set up grasscutter, poultry, rabbit and other small livestock businesses.

The many training videos on small livestock, intercropping with legumes and rural food processing offer viable alternatives to the hunting for bush meat. These enterprises may eventually prove more effective in reducing bush fires than lecturing rural people about their adverse environmental impacts. Positive solutions that focus on the benefits of biodiversity are a better way to promote conservation.*

Related videos

The videos cited in this story, including one on grasscutters, or cane rats, which can be raised on family farms, are:

AMEDD (2017) *Making a Condiment from Soya Beans*. Hosted by Access Agriculture. 10 minutes.
www.accessagriculture.org/making-condiment-soya-beans

DEDRAS (2018) *Growing Annual Crops in Cashew Orchards*. Hosted by Access Agriculture. 9 minutes.
www.accessagriculture.org/growing-annual-crops-cashew-orchards

DEDRAS (2020) *Making Soya Cheese*. Hosted by Access Agriculture. 9 minutes.
www.accessagriculture.org/making-soya-cheese

DEDRAS, Nawaya and Songhaï Centre (2016) *Feeding Grasscutters*. Hosted by Access Agriculture. 9 minutes.
www.accessagriculture.org/feeding-grasscutters

A version of the section above was previously published on www.agroinsight.com in 2018, by Paul Van Mele.

Forgotten Vegetables

When Johan Hons and Vera Kuijpers started their farm *Het Eikelenhof* (garden full of acorns) in 1989 in the eastern province of Limburg in Belgium, organic food was not yet in fashion. Uncertain about the market and with little knowledge available on organic farming, the young couple was convinced that they could make it work. And they did.

Continuous learning about crops, varieties, soil and cropping practices has already allowed the family to live from the farm for more than 30 years.

One of their marketing innovations is the farm shop, which the couple started in 2010 and which Vera enthusiastically opens on Fridays and Saturdays. On one of my visits, a customer asked Vera where she got rhubarb (*Rheum rhabarbarum*) so early in the spring. Vera told the following story:

> Around 2000, Johan was driving to the north of the Netherlands, when he happened to see a field of rhubarb with plants that were well ahead of his own. Johan's rhubarb had barely started to come up. He made a mental note, and on his way back decided to pay a visit to the farmer. To his disappointment the farmer said something like, 'There is no way I am going to give you even a single plant.' It was his way of keeping a monopoly on this special, early variety.
>
> Back at home in Flanders, Johan shared his disappointing experience with Vera and they got on with their farming. But in 2005, Johan decided to try his luck a second time. On a trip to the Netherlands, he stopped at the same house. The farmer was really old by now and to Johan's surprise said, 'You can have my rhubarb, but only under one condition: if you take them all.' The farmer was retiring and wanted to make sure that his rhubarb variety lived on. But the Dutch farmer had three acres (over a hectare) of the red vegetable, a lot of work to dig up and transplant. But Johan rose to the challenge.
>
> Having made up his mind, Johan drove back home, gathered some friends, put his tractor on a truck and began uprooting the large field of rhubarb which he then planted on their own land.

Vera proudly ended the story by saying, 'We now have the earliest rhubarb in Belgium and supermarkets are lining up to buy our first harvest.'

When her customer asks if she could get a few plants for her own garden, Vera smiles and says, 'You can try to convince Johan, but I think you may have to wait another 20 years or so.'

Rhubarb has been grown for more than 5000 years and was first cultivated by the Chinese (who also gave us rice, soy bean and oranges). They used the root as a medicine. Rhubarb only became popular as a vegetable in Britain in the late 18th century (following the trade with China) and has only been cultivated in Belgium since around 1900. People nearly stopped eating the sour stalks during the Second World War, possibly because sugar was so severely rationed. Society's tastes in food do change with time, and old fruits and vegetables have been coming back into vogue these last few years.

Farmers like Johan and Vera are among the thousands of committed keepers of the biodiversity of our planet. By cultivating forgotten fruits and vegetables, family farmers are as important as the large genebanks maintained by research agencies around the world. Having a special kind of vegetable can help give a farm family a marketing niche, and make it profitable to preserve the variety. Biodiversity is about having your pie, and eating it too.*

📹 **Related video**

Farmers in tropical countries also nurture relatively forgotten vegetables, like okra. Smallholders in Mali explain how to harvest and store okra.

AMEDD (2022) *Harvesting and Storing Okra*. Hosted by Access Agriculture. 12 minutes. www.accessagriculture.org/harvesting-and-storing-okra

---

*A version of the section above was previously published on www.agroinsight.com in 2015, by Paul Van Mele.

References

Dong, J., Gruda, N., Li, X., Cai, Z., Zhang, L. *et al.* (2022) Global vegetable supply towards sustainable food production and a healthy diet. *Journal of Cleaner Production* 369, 133212.

IPES-Food (2016) *From Uniformity to Diversity: A Paradigm Shift from Industrial Agriculture to Diversified Agroecological Systems*. International Panel of Experts on Sustainable Food Systems. Available at: https://ipes-food.org/report/from-uniformity-to-diversity/ (accessed 16 October 2024).

Juroszek, P., Lumpkin, T., Easdown, W. and Tsai, H.H. (2008) Organic vegetable production research needs in the tropics and subtropics. In: Köpke, U. (ed.) *Organic Agriculture in the Tropics and Subtropics*. ISOFAR, Bonn, Germany, pp. 29–51.

Montgomery, D.R. (2017) *Growing a Revolution: Bringing Our Soils Back to Life*. Norton, New York.

Pinker, S. (2018) *Enlightenment Now: The Case for Reason, Science, Humanism and Progress*. Penguin Books, London.

Quispe, M., Baldiviezo, E. and Laura, S. (2008) *Yapuchiris: Ofertantes Locales de Servicios de Asistencia Técnica (Yapuchiris: Local Technical Assistance Service Providers)*. PROSUCO, Cooperación Suiza, La Paz, Bolivia.

Quispe, M., Laura, S. and Baldiviezo, E. (2018) *Yapuchiris: Un Legado para Afrontar los Impactos del Cambio Climático (Yapuchiris: A Legacy to Face the Impacts of Climate Change)*. PROSUCO, Cooperación Suiza, La Paz, Bolivia.

Toensmeier, E. (2016) *The Carbon Farming Solution: A Global Toolkit of Perennial Crops and Regenerative Agriculture Practices for Climate Change Mitigation and Food Security*. Chelsea Green Publishing, White River Junction, Vermont.

5   Local Seed

Abstract

Local seed and local crop varieties are important for biodiversity and food security, in spite of policies that favour commercial seeds. In Malawi and Guatemala, farmer seed banks and seed fairs help local people exchange and preserve their cherished crop varieties. A study in Kenya finds good germination rates for informally produced African leafy vegetable seeds. Farmers in Ghana, like Issah Bukari, innovate by growing groundnuts alongside other crops to ensure a steady supply of seed. In Peru, the Association of Guardians of Native Potatoes (AGUAPAN – *Asociación de los Guardianes de la Papa Nativa del Centro del Perú*) fights to keep native potato varieties alive through direct marketing and consumer education.

Richness in Diversity

For decades, new crop varieties have been bred by relying heavily on farmers' knowledge and the local landraces they grow. Landraces have provided a major gene pool readily used by breeders to breed crops that are better adapted to drought, floods, pests and diseases. But with increased pressure from the private sector and insufficient support from the public sector, many rural communities struggle to maintain their diversity of crops and food, as I recently learned in Malawi.

When my wife, Marcella, who is also the cameraperson at Agro-Insight, and I were asked by the Global Forum on Agricultural Research and Innovation (GFAR) to make a video on farmers' rights to seed, we only had a faint idea of how strongly the debate raged among development organizations, policy makers and farmers. We were surprised to learn that in Malawi, a draft seed policy had been written that would force farmers to buy commercial seed only.

Driving from the capital city of Lilongwe to the northern town of Rumphi, we passed many fields with dried stubble, where maize and tobacco had been recently harvested. Beyond these bleak fields, the rolling, dusty hills were decorated with trees flowering in white, orange, purple and other

© Jeffery W. Bentley and Paul Van Mele 2025. *Agroecology in Practice: From Local Initiatives to Global Scaling Through Video* (J.W. Bentley and P. Van Mele) DOI: 10.1079/9781800628793.0005

colours. But the beauty of the landscape doesn't stop one from seeing the dire poverty in which the people live.

Maize is the staple food and many farmers grow it as a cash crop, encouraged by government fertilizer subsidies. Farmers who accept the subsidy are obliged to plant only hybrid maize seed.

Families growing tobacco for multinational companies have basically sacrificed their lives to the crop, but unlike the fluctuating world market price for tobacco over the years, their living conditions have remained stubbornly low at all times.

The reliance on these two key crops is beginning to change. Recent development efforts have started to take crop and food diversification seriously. As I talked to farmers over the next few days, it dawned on me how much effort is required for farmers to preserve local crop varieties that have been nurtured over many years. Many families have abandoned their traditional crops and dishes, and the current generation of farmers has little idea of how to grow anything else apart from maize and tobacco.

On our first day of filming, we visited the community seed bank in Mkombezi (Fig. 5.1). As member farmers arrived in small groups, we filmed the shelves lined with glass jars full of seeds of local varieties of sorghum,

Fig. 5.1. Community seed banks in Malawi often store tonnes of seed of local and indigenous varieties.

millet, maize, beans, groundnuts and Bambara groundnuts. Shadreck Kapira, secretary of the seed bank, proudly explained:

> We keep seed of our local varieties and multiply them to share with our members, and also to supply non-members. At this moment we have 14 t of seed in our store room.

Outside the seed bank more farmers had gathered. With the support of a local NGO, some eight farmers from southern Malawi had travelled over 600 km to meet fellow farmers in the north. The next day, they all attended a seed fair to exchange and sell seed of their food crops. The visiting farmers proudly displayed small plastic bags, each containing precious seeds. Each lot is poured onto a red, blue or green plastic plate and a label attached with the name of the farmer and seed variety.

During the group discussion the farmers from the north showed great interest in the sorghum varieties on offer from their colleagues from the south. With the changing climate the hybrid maize varieties do not perform as well as they used to. If rains are poor, a farmer risks losing her entire crop. Some of the local sorghum varieties mature in just 2 months, a month earlier than the hybrid maize. And the local varieties better withstand drought.

Farmers also talked about how they use different crops to prepare food and drinks for special events, such as weddings or the nomination of village chiefs. Millet is one of their favourite crops. It produces a porridge which is not only more nutritious than that made from maize, but can be prepared with less water and without cooking, so there is no need for fuel wood. Millet is also an essential ingredient of traditional sweet and sour beers.

The next morning, we left early, just before dawn at 5 o'clock, to reach Mpherembe on time for a seed and food fair. The local community had fenced off an area near the water well and tied bundles of local grasses to sturdy poles to keep out the dust-laden wind (Fig. 5.2). Local NGO staff registered each farmer and the type and amount of seed they brought to the fair. Women had also prepared a diverse range of foods, and when I peeked under the lid of the occasional plastic bucket, I found millet beer, an important part of a fair.

According to the International Treaty on Plant Genetic Resources for Food and Agriculture, signed by over 140 governments across the world, every farmer has the right to exchange and sell their local seed. When we interview one of the farmers, Bena Phiri, she is explicit:

> My rights, I can say that my local crops that I grow are mine and no one can have control over my seed. I have the right to sell them at my own will and no one can say anything because they are mine.

However, in 2017 a draft seed policy wanted to discourage farmers from saving their own seed. If this policy had been approved, it would have spelled disaster for local crop varieties. Most agro-dealers have few varieties for sale, and hardly any are local. The stores mainly sell hybrid maize from Monsanto, Syngenta and perhaps one or two other multinational companies. Fortunately, under pressure from development agencies and farmer

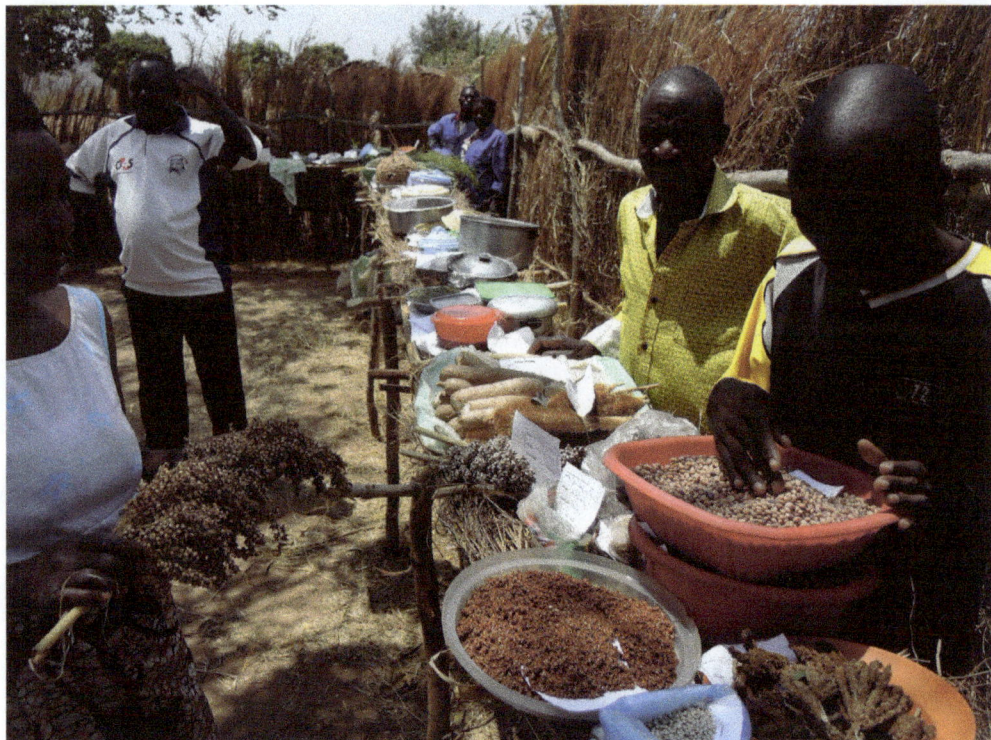

Fig. 5.2. Seed fairs allow farmers to exchange seed and knowledge.

organizations, the draft seed policy was modified. The final seed policy recognized that the informal seed sector, especially farmer-saved seed, was the main source of seed in Malawi (Malawi Government, 2018). Language that discouraged the 'recycling of self-pollinated seeds beyond their genetically recommended recycling times' was modified to explicitly state that 'over recycling' of seed would be discouraged by 'creating awareness', rather than through supervision by the National Seed Commission (Malawi Government, 2018). The changes in the policy were a partial victory for the farmers: their rights to seed were not explicitly protected, but the importance of the informal seed system was acknowledged and saving seed was not penalized.

We hope that our video on farmers' right to seed, available in English and two Malawian languages (Chichewa and Tumbuka), and various other African languages, will help to raise farmers' awareness across the country and beyond. Distributed by Access Agriculture and its diverse partners in Malawi, the videos have been shown in farm clubs and on local TV and aired on the radio. Many farmers also obtained copies to view the video directly on their inexpensive mobile phones.

It is ironic that wealthy people are now able to access more food diversity than ever, at a time when the poor could have many of their local crop varieties wiped out by misguided laws. The media has a role to play in raising awareness among farmers, legislators and consumers and to help maintain local cultures based on a rich diversity of crops and foods.*

📹 **Related video**

The experience narrated above is captured in the following video, available in 14 languages.

Agro-Insight (2017) *Farmers' Rights to Seed: Experiences from Malawi*. Hosted by Access Agriculture. 16 minutes.
www.accessagriculture.org/farmers-rights-seed-experiences-malawi

---

*A version of the section above was previously published on www.agroinsight.com in 2017, by Paul Van Mele.

Homegrown Seed Can Be Good Seed

African leafy vegetables are important for nutrition and are increasingly for sale. Shops in Africa are now starting to sell seed packets of leafy vegetables to farmers and gardeners. But seed produced by farmers is also sold informally, in small town markets.

A study in Kenya (Croft *et al.*, 2018) suggested that this informal seed could be fairly good. Marcia Croft and colleagues compared 24 lots of seed for two kinds of African leafy vegetables: amaranth (*Amaranthus* spp.) and nightshade (*Solanum* spp.). For each kind of seed, the study compared six lots of informal seed with six lots of formal seed. Since few companies in Kenya sell vegetable seed, the six lots of formal seed were made up of one lot from a Kenyan seed company and five lots from Tanzania, from the World Vegetable Center (WorldVeg), formerly the Asian Vegetable Research and Development Center (AVRDC). The germination rates for the informal seed were acceptable: 59% for amaranth and 84% for nightshade, while less than 30% of the formal seed sprouted. The article does not explain why the formal seed had such an abysmal germination rate. Perhaps in future studies the formal seed will perform better.

The study raises questions: How good is vegetable seed in other African countries? And, how fresh was the seed in the shop and in the AVRDC collection? Future studies should clearly separate commercial seed from the collection of a research centre.

The Kenyan seed customers themselves form two distinct groups. The study showed that the customers of African leafy vegetable seed include poorer farmers (who have less land to grow their own seed), men and families with a salaried wage earner, i.e. weekend gardeners who don't have time to produce seed.

Smallholders and gardeners could demand more seed in the future. To supply them, the study concludes that the informal seed markets should be strengthened, rather than supporting formal market development for African leafy vegetable seed. Good seed, produced locally, will benefit the poor and gardeners trying to grow healthy food for their families.*

▣◀ Related video

Farmers in West Africa explain how to plant okra from seed.

Agbangla, A. (2020) *Making a Good Okra Seeding*. Hosted by Access Agriculture. 12 minutes.
www.accessagriculture.org/making-good-okra-seeding

---

*A version of the section above was previously published on www.agroinsight.com in 2017, by Jeff Bentley.

Meeting the Need for Groundnut Seed

While saving seed on-farm is common practice, many farmers also rely on other seed sources, like local markets. When the rains start, however, the demand for seed may be so high that the market fails to supply it in a timely way. Fortunately, smallholders have their own creative ways of meeting the need for seed, as we recently found out.

On our first day of filming in Bawku, a small town in northern Ghana separated from Burkina Faso by a river, we were amazed by farmers' ingenuity and drive to succeed. We were making farmer training videos on onions under a scorching sun, in 40°C heat. Issah Bukari, like many other farmers, grows onions in the dry season because that is when onions are attacked by fewer diseases.

'Let me show you where we get our water to irrigate our crops', Issah insisted, leading us to the nearby river.

To our surprise the river had no water. Women carried their babies on their backs and pushed their bicycles across the dry riverbed. Every year, the border between the countries changes from a wide river to a winding, sandy avenue.

Issah showed us a 10-m-wide circular pit that he hired someone to dig for him. We began to realize how much effort it took to get water in the dry season.

'I pay a lot for people to dig the well. Every 3 days they have to dig deeper to get enough water. But it is worth it, because in the dry season water is gold.'

Issah grew his onions in sunken beds to conserve the precious water. Issah's garden was quite diverse. Papaya, mango and cashew trees grew scattered across his fields. On the edges of some of the sunken beds, healthy maize was grown, not as a cereal crop, but as a snack: roasted maize cobs fetch a good price this time of the year. We also see some hibiscus plants, the flowers of which are used to make a refreshing red lemonade.

What looked like creeping weeds from afar, the only green spots left after the onion harvest, turned out to be groundnut plants.

'How come you grow some scattered groundnuts in the dry season?' I asked Issah.

'This is my guarantee to have groundnut seed by the time the rains come', he replies.

By growing groundnuts for seed on the edge of the sunken onion beds, the groundnuts profit from the irrigation water given to the onions.

Groundnut seed, like many other legume crops, is rich in oil and quickly gets rancid in warm weather. This makes it difficult for farmers to save groundnut seed for the next season. As illustrated in *African Seed Enterprises* (Van Mele *et al.*, 2011), farmer groups and private seed companies that provide legume seed in one season may struggle to guarantee a supply in the next.

Continuity of supply is a big challenge. Sometimes seed enterprises cannot get enough foundation seed themselves to grow enough certified legume seed for when farmers need it (except when large orders are placed in advance by development projects). To deal with the uncertainties of access to water and access to seed, farmers like Issah Bukari innovate with cropping patterns, space and time.

Seed agencies are often highly critical of farmers' practices to save seed, claiming that farmer-saved seed does not even deserve to be called seed at all. Of course, the seed companies have a vested interest in destroying local seed production. As we see here, there is a lot more to saving seed than simply putting it in a sack (or 'brown bagging' as the practice is sometimes disapprovingly called). Farmers use creative solutions, sometimes going to great efforts to produce small amounts of the varieties they need in the dry season to have seed ready to meet the rains.*

◼ Related video

This video is part of a series on how to produce onions commercially on small farms.

Agro-Insight (2016) *How to Make a Fertile Soil for Onions*. Hosted by Access Agriculture. 7 minutes.
www.accessagriculture.org/how-make-fertile-soil-onions

A version of the section above was previously published on www.agroinsight.com in 2016, by Paul Van Mele.

Native Potatoes

Peru's native potatoes are a living treasure of 4000 varieties that come in red, purple, yellow and black (see CIP, 2023). Round or long, smooth or knobby, each one is different, and tasty. But for years city people ignored the native potato, considered to be the inferior food of poor people (Horton *et al.*, 2022).

Some farmers and their allies are fighting to keep the native potato alive. Over 20 years ago, Peruvian agronomist Raúl Ccanto was one of the people who realized that native potatoes could survive, if people in the city would buy them.

A brand name was created, Mishki Papa (tasty potato), and the little tubers were displayed in a net bag so customers could see their unique beauty. To produce the potatoes, 50 farmers were organized into the newly created Association of the Guardians of Native Potatoes of Peru (AGUAPAN).

I told Raúl that I used to buy these potatoes at an upscale supermarket when I lived in Peru in 2010. Raúl explained that AGUAPAN was no longer selling through the supermarket, which would only pay for the potatoes 2 weeks after they had taken delivery and would return any unsold ones, paying the farmers only 1.30 soles (about US$0.30), while charging customers 4.30 soles. Perhaps most discouraging, the supermarket only accepted three or four varieties of potatoes, while farmers grew dozens. As Raúl explained, if consumers only bought four varieties, the others would still be endangered.

In recent years, a farmers' cooperative headquartered in the Netherlands (Agrico) and a company based in the Netherlands (HZPC) have each given AGUAPAN 15,000 Euros to help them market potatoes. While making a farmer learning video with the team from Agro-Insight, we visited the president of AGUAPAN, Elmer Chávez, while he harvested native potatoes with his family in the village of Vista Alegre, in Huancavelica, Peru, at 3900 m above sea level (12,800 ft). At this staggering altitude, where we struggled just to breathe and walk, the Chávez family was hard at work, carefully unearthing each variety (Figs 5.3 and 5.4).

The family grows 80 varieties of potatoes. During the harvest, they select five tubers of each variety as seed for next year. The rest are to eat at home and to sell. The family works hard against a deadline. We were there on a Friday, and on Monday morning don Elmer had to be at a trucking company in Huancayo, 30 km away, to ship half a tonne of potatoes.

In Lima, representatives of Yanapai (an NGO that collaborates with AGUAPAN) will receive the potatoes, advertise them on social media, keep them in a warehouse and take orders from individual customers. On the following Friday, the potatoes will be sold in 2-kg net bags, with as many as 18 varieties in each little sack. Raúl explains that this is called a *chaqru* (from the Quechua word for 'mix'). Each farm family produces its own special mix, selected over the years to have the same cooking time and to combine nicely on the plate.

To promote the potatoes, Yanapai has made a catalogue of the varieties and a booklet describing individual farmers and the unique mix of potatoes that each one has (MINAGRI *et al.*, 2017).

As agronomist Edgar Olivera of Yanapai explains, the delivery service still requires some financial and technical support, but the hope is that one

Fig. 5.3. The Chávez family saves five tubers of each potato variety to use as seed the following year.

day it will be self-sustaining. Many farmers have grown children who now live in the capital city of Lima. Some of the children of farmers may one day be able to earn money selling the native potatoes from their home villages, turning the gem-like potatoes of their parents into a real source of income for the families who nurture them.*

📹 **Related video**

The experiences described in this section are also captured in the following video.

Agro-Insight and Yanapai (2022) *Recovering Native Potatoes*. Hosted by Access Agriculture. 16 minutes.
www.accessagriculture.org/recovering-native-potatoes

A version of the section above was previously published on www.agroinsight.com in 2022, by Jeff Bentley.

Fig. 5.4. Native potatoes, in sizes and colours not seen in every supermarket.

Seed Fairs

Seed fairs are gaining in popularity around the world and are a great way to encourage farmers and gardeners to conserve global biodiversity. But the fairs can do more than just provide an opportunity for people to exchange and sell seed, as I recently learned during a visit to Guatemala to make a farmer training video on farmers' rights to seed, with a particular focus on women in biodiversity management. In Guatemala, donor agencies and organizations have supported community biodiversity conservation initiatives for over a decade.

Our local partner, the Association of Cooperatives of Huehuetenango (ASOCUCH – *Asociación de Cooperativas de Huehuetenango*), is an umbrella organization of 20 cooperatives and farmer associations, representing some 9000 farm families in the western highlands of Huehuetenango. On Sunday, one day before the actual seed fair starts, we visited the venue. The seed fair has become a large annual event, unlike in Malawi, where seed fairs are more sporadic. The fair attracts hundreds of people from across the highlands, some travelling long distances. One elderly woman told me she rode a bus for 5h to get there.

The seed fair is a lively social event, with a Ferris wheel and stalls with games. Other stalls offer sweets and nuts. There are artistically carved wooden horses with leather saddles where people can sit and have their photo taken against a lush, painted background of mountains and waterfalls. A young boy sells mandarin oranges from a wheelbarrow he gently pushes through the crowd, while indigenous women sell traditional delicacies. Families with grandparents and kids relish the event as the region does not have such a large fair very often.

But there is more to the fair than having fun and eating. The seed fair is held on school grounds, and I soon see farmers in intricately woven traditional clothes lining up to register for classes. There are four large rooms where farmers can learn about potato, agrobiodiversity, climate change and women's rights. My wife, Marcella, who is the cameraperson at Agro-Insight, and I first attend the talks in the agrobiodiversity room, where Juanita Chaves from GFAR explains about farmers' rights to seed. To my surprise, this is followed by two presentations on aflatoxins in maize by staff from a local NGO. The presenters graphically explain the relation between mouldy maize cobs and the disfigurement of children and internal organs. As most farmers store their own maize seed, they need to be aware of the risks of fungal infections. I am still a little puzzled as to how this relates to the seed fair and agrobiodiversity conservation, but after lunch all becomes clear.

We accompany the farmers who attended the aflatoxin sessions to the Clementoro Community Seed Bank, less than 10 km away. The farmers see seeds stored in plastic jars, clearly labelled and neatly stacked on the shelves. In the middle of the room, a young agricultural graduate, who works at the seed bank, shows farmers how they can detect if their seed is contaminated with aflatoxins by using a simple methanol test. 'When you store your maize crop and seed, you need to be sure it has less than 13% moisture so that moulds will not develop', the enthusiastic young woman explains. 'Here at the seed bank, you can have your seed tested and conserved in optimal conditions', she continues.

Seed is one of farmers' most precious resources, and storing it at a community seed bank requires lots of trust. Farmers need to know that their seed will be safely stored until they need it, for the next growing season or even a few years later, whenever the need arises. By organizing seed fairs, seminars and visits to community seed banks, ASOCUCH is building trust through sharing knowledge and explaining clearly what they do.

The next day, we film the seed fair itself. There is an overwhelming abundance of crop varieties, fruits, medicinal plants and even some ornamental ones. Farmers and their families are clearly excited as seed and planting material changes hands. There is brisk trading between farmers. While some exchange materials, most buy and sell seed. People tell each other about the seeds they have on offer. ASOCUCH, with the support of GFAR, has also prepared a booklet with traditional recipes. Copies are spread on tables at the entrance and they go like hot cakes.

There is a judging competition to find the best seeds. Judges visit each stand, measuring maize cobs, counting seeds, weighing potato seed tubers and taking notes. Agrobiodiversity is a serious matter. At the same time, outside the schoolhouse, sheep are being rated by another set of judges.

In the late afternoon, the results are shared with the audience. People had brought dozens of varieties and over a thousand accessions of various crops. The audience is excited, and so are we. This had been a fascinating two-day event, and the drive of the farmers and their organizations made us hopeful for the future. Local initiatives are where conservation begins, but they need the support of local authorities, governments and international organizations to increase their impact.

Everyone had a good time. More importantly, farmers made new contacts, acquiring seeds of traditional varieties that may have been lost in some areas and introducing them to new areas. The farmers learned about saving seed, and especially that local people have certain rights to seeds – they can plant their own native varieties as they wish, for example – and that these rights matter hugely in sustaining local agriculture.*

📹 Related video

Here is the video about community seed banks in Guatemala.

Agro-Insight (2018) *Video Farmers' Rights to Seed: Experiences from Guatemala*. Hosted by Access Agriculture. 14 minutes.
www.accessagriculture.org/farmers-rights-seed-experiences-guatemala

---

A version of the section above was previously published on www.agroinsight.com in 2018, by Paul Van Mele.

References

CIP (2023) Native potato varieties. International Potato Center (CIP), Lima, Peru. Available at: https://cipotato.org/potato/native-potato-varieties/ (accessed 16 October 2024).

Croft, M.M., Marshall, M.I., Odendo, M., Ndinya, C., Ondego, N.N. *et al.* (2018) Formal and informal seed systems in Kenya: Supporting indigenous vegetable seed quality. *The Journal of Development Studies* 54, 758–775.

Horton, D., Devaux, A., Bernet, A., Mayanja, A., Ordinola, M. *et al.* (2022) Inclusive innovation in agricultural value chains: Lessons from use of a systems approach in diverse settings. *Innovation and Development* 13, 517–539.

Malawi Government (2018) *National Seed Policy*. Department of Agricultural Research Services, Ministry of Agriculture, Irrigation and Water Development, Lilongwe, Malawi.

MINAGRI (Ministerio de Agricultura y Riego), Grupo Yanapai, INIA (Instituto Nacional de Innovación Agraria) and CIP (Centro Internacional de la Papa) (2017) *Catálogo de Variedades de Papa Nativa del Sureste del Departamento de Junín - Perú (Catalogue of Native Potato Varieties from the Southeast of the Department of Junín - Peru)*. CIP, Lima, Peru. Available at: https://cgspace.cgiar.org/bitstreams/b9a6e152-e828-4ba3-b13f-9f7162ba1762/download (accessed 16 October 2024).

Van Mele, P., Bentley, J.W. and Guéi, R.G. (eds) (2011) *African Seed Enterprises: Sowing the Seeds of Food Security*. CAB International, Wallingford, UK.

6   Healthy Soils

Abstract

Techniques like soil conservation, cover crops, composting and mulching can revive degraded farm soils, stop erosion and manage weeds. Crop rotation combats nutrient losses and crop diseases. Researchers can play a crucial role to help local communities to conserve native plants that stop soil erosion, or to promote, multiply and distribute beneficial soil microorganisms.

Local companies can promote the use of earthworms, while farmer-to-farmer videos help to share knowledge about homemade solutions for enhancing soil health, and pave the way for a more sustainable and productive agricultural future.

Living Soil: A Film Review

In the opening scenes of the film *Living Soil*, we see the Dust Bowl: the devastated farmland of the 1930s in the southern plains of the USA. Between 30 and 50 years of ploughing had destroyed the soil, and in times of drought, it drifted like snow.

As the rest of this 1-hour film shows, there is now some room for optimism. Nebraska farmer Keith Berns starts by telling us that most people don't understand the soil, not even farmers. But this is changing as more and more farmers, large and small, organic and conventional, begin to pay attention to soil health, and to the beneficial microbes that add fertility to the soil. Plants produce carbon and exchange it with fungi and bacteria for nutrients.

Mimo Davis and Miranda Duschack have a 1-acre (0.4-ha) city farm in Saint Louis, Missouri. The plot used to be covered in houses, and it was a jumble of brick and clay when the urban farmers took it over. They trucked in soil, but it was of poor fertility. So, they rebuilt the soil with compost and cover crops, like daikon radishes. Now Davis and Duschack are successful farmer-florists – growing flowers without pesticides, so customers can bury their noses in a healthy bouquet.

A few scientists also appear in the film. Kristin Veum, a soil scientist with the US Department of Agriculture (USDA), says that most people know that

© Jeffery W. Bentley and Paul Van Mele 2025. *Agroecology in Practice: From Local Initiatives to Global Scaling Through Video* (J.W. Bentley and P. Van Mele)
DOI: 10.1079/9781800628793.0006

legumes fix nitrogen, but few know that it's the microbes in association with the plants' roots that actually capture the nitrogen from the air.

Indiana farmer Dan DeSutter explains that mulch is important not just to retain moisture, but also to keep the soil cool in the summer. This helps the living organisms in the soil to stay more active. Just like people, good microbes prefer a temperature of 20–25°C. When it gets either too hot or too cold, the microorganisms become less active. Cover crops are also important, explains DeSutter, 'Nature abhors a mono-crop.' DeSutter plants cover crops with a mix of 3–13 different plants, and this not only improves the soil, but keeps his cash crops healthier.

Nebraska's Keith Berns plants a commercial sunflower crop in a mulch of triticale straw, with a cover crop of Austrian winter pea, cowpeas, buckwheat, flax, squash and other plants growing beneath the sunflowers. This diversity then adds 15 or 20 bushels per acre of yield (1–1.35 t per hectare) to the following maize crop. Three rotations per year (triticale, sunflower and maize), with cover crops, build the soil up, while a simple maize–soy bean rotation depletes it.

Adding carbon to the soil is crucial, says DeSutter, because carbon is the basis of life in the soil. In Indiana, half of this soil carbon has been lost in just 150–200 years of farming, including 50 years of intensive agriculture. No-till farming reduces fertilizer and herbicide costs, increases yield and the soil improves: a win-win-win. This also reduces pollution from agrochemical runoff.

As Keith Berns explains, the Holy Grail of soil health has been no-till without herbicides. It's difficult to do, because you have to kill the cover crop to plant your next crop. One option is to flatten the cover crop with rollers, and another solution is to graze livestock on the cover crop, although he admits that it's 'really hard' to get this combination just right.

USDA soil health expert Barry Fisher says, 'Never have I seen among farmers such a broad quest for knowledge as I'm seeing now.' The farmers are willing to share their best-kept secrets with each other, which you wouldn't see in many other businesses.

Many of these farmers are experimenting largely on their own, but a little state support can make a huge difference. In the 1990s in Maryland, the Chesapeake Bay had an outbreak of Pfiesteria, a disease that was killing the shellfish. Scientists traced the problem to phosphorous from chemical fertilizer runoff. Maryland's State Government began to subsidize and promote cover crops, which farmers widely adopted. After 20 years, as Chesapeake Bay waterman James 'Ooker' Eskridge explains, the bay is doing better. The sea grass is coming back. The blue crab population is doing well, the oysters are back and the bay looks healthier than it has in years.

Innovative farmers, who network and encourage each other, are revolutionizing American farming. As of 2017, US farmers had adopted cover crops and other soil health measures on at least 17 million acres (6.9 million hectares), a dramatic increase over 10 years earlier, but still less than 10% of the country's farmland. Fortunately, triggered by increased consumer awareness, these beneficial practices are catching on, which is important, because healthier soil removes carbon from the atmosphere, reduces agrochemical use, retains moisture to produce a crop in dry years and grows more food. The way forward is clear. Measures like targeted subsidies to help farmers buy seed of cover crops have been instrumental to help spread agroecological practices. Experimenting farmers must be supported with more public research and with policies that promote healthy practices like mulching, compost, crop rotation and cover crops.*

Related video

This is the video *Living Soil*, reviewed in this section.

Wright, C. (2018) *Living Soil*. Soil Health Institute. Hosted by EcoAgTube. 60 minutes. https://ecoagtube.org/content/living-soil

---

*A version of the section above was previously published on www.agroinsight.com in 2020, by Jeff Bentley and Paul Van Mele.

Stop Erosion

When farmers have limited access to arable land, and soils are poor, limiting soil erosion can make the difference between harvesting a crop or nothing at all.

Soils are indeed at the core of any crop production system. Without a healthy soil, crops cannot thrive. While measuring the effect of soil erosion at national and global scales is near impossible, all farmers see the difference when effective soil conservation techniques are in place. Controlling erosion is becoming increasingly urgent as climate change is leading to rains falling more erratically and intensely than before.

From the gentle rolling lands in Burkina Faso to the steep hills in northern Vietnam, I have seen the devastating effects of rainfall on poorly managed soils. Even on slopes as gentle as 5°, the torrential tropical rains wash away the top soil and seal the top layer, after which no more water can penetrate the soil. To remedy this, farmers in Burkina Faso learned about making contour bunds (raised ridges every 20 m across the field) to allow the rainwater to infiltrate (Fig. 6.1). On steeper slopes, vegetation barriers or terraces can stop soils from eroding (Fig. 6.2).

Depending on the slope, type of soil, availability of labour and other resources, a wide range of options are available to improve soil and water management. Networks such as the World Overview of Conservation Approaches and Technologies (WOCAT) support organizations working on the ground with farmers by making hundreds of sustainable soil and water management technologies available on an authoritative website.

While many development agencies and projects believe that encouraging smallholder farmers to use synthetic fertilizers is the quickest way to solve

Fig. 6.1. Young farmers in Burkina Faso make contour bunds to stop soil erosion.

Fig. 6.2. On sloping land in northern Vietnam, fodder grasses are grown in between cassava to conserve soil.

low crop productivity, without proper soil conservation techniques, farmers will see most of their cash investment wash down the drain.*

■◀ Related videos

The videos cited in this story are:

Agro-Insight (2012) *Contour Bunds*. Hosted by Access Agriculture in about 40 languages. 15 minutes.
www.accessagriculture.org/contour-bunds

Agro-Insight (2015) *Grass Strips against Soil Erosion*. Hosted by Access Agriculture in more than 20 languages. 12 minutes.
www.accessagriculture.org/grass-strips-against-soil-erosion

---

*A version of the section above was previously published on www.agroinsight.com in 2017, by Paul Van Mele.

Awakening the Seeds

In much of the Bolivian Altiplano, the native vegetation has been largely stripped away for firewood and quinoa-growing, causing wind-blown soil erosion. Fortunately, most rural Bolivians are now using bottled natural gas for cooking, so firewood cutting has nearly stopped. But conserving the soil will require several tactics, including windbreaks and encouraging the recovery of native plant cover. In this book we have stressed how agroecology requires support from research. This section describes how a few determined Bolivian researchers are learning to replant the vegetation and to germinate the seeds of native plants to conserve the soil.

These Andean high plains were once covered by scrubland, comprising low-lying bushes, needle grasses and other hardy plants well adapted to the harsh conditions. Llamas foraged on this waist-high forest without damaging it. But as more land was ploughed up for quinoa, and more of the bushes were cut for firewood, the native vegetation started to vanish (Bonifacio *et al.*, 2022).

Rural families in this part of Bolivia used to make long, narrow stacks of dried brush. But the bushes are now mostly gone, and so are the stacks of firewood. Fortunately, explains plant researcher Dr Alejandro Bonifacio, people are now cooking with bottled natural gas, so they don't need to uproot brush for firewood, but this respite has come too late. In many places, the deforestation has been so complete that there are no seed-bearing plants left to provide for natural regeneration. So, Dr Bonifacio and his team travel around the Altiplano, collecting seed of different shrubs, planting the seed in nurseries and then taking the seedlings to sympathetic farmers who are interested in restoring the dry plains (Bonifacio *et al.*, 2022).

Seeds of wild plants will seldom germinate if simply scattered on the ground. The plants are adapted to harsh environments, and the seed enters dormancy, only to be awakened by some specific environmental signal.

Bonifacio and his students study each plant to determine what will break its dormancy. For example, the *k'awchi* (*Suaeda foliosa*), a small woody plant, is so adapted to this land of high winds and rocky soil that its tiny seed must be tumbled over the rough ground and 'scarified' before it will germinate. Bonifacio and his team have also learned that it can be scarified by rubbing it in sand or by putting it in a weak solution of sodium hypochlorite for 20 minutes.

On the arid Altiplano, much of the native vegetation is cactus, some of it bearing delicious fruit. In a boutique restaurant in the big city of La Paz, Bonifacio was shocked to hear that the chef was asking him for a supply of one native cactus, called *achakana* (*Neowerdemannia vorwerckii*). Yes, achakana is edible, but it takes many years to grow to the size of a tennis ball. The Aymara people used to eat the cactus as famine food when the crops failed, but achakana could be driven to extinction if it starts to be served up in the fashionable eateries of La Paz. So, Bonifacio taught himself how to propagate it.

It was tricky. At first, the seed failed to germinate. Bonifacio learned that as the fruit matures the seed goes into a deep dormancy. Then one day, by serendipity, Bonifacio discovered a little bag of fruit that had been harvested green and then forgotten. When he opened the rotting fruit, he found that all of the seeds were germinating. He proudly showed me a small, 3-year-old plant that he had grown from seed.

The *pasakana* (*Trichocereus pasacana*) is another endangered cactus that grows so tall that the Andean people once used its ribs to roof their houses. The fruit is also delicious, yet getting the seed to germinate was impossible. Then Bonifacio found that the pasakana seed would germinate if it was taken from immature fruit. With the help of a student, he now has 1200 little pasakana plants, all in demand from a municipal government in Oruro which wants to plant them out.

More people than ever want to grow native plants for fruit, fodder and soil conservation, but each species has its own unique requirements for coming to life. Fortunately, there are patient researchers working to unlock these mysteries and come up with practical recommendations that can help restore degraded lands.*

🎥 Related video

Dr Bonifacio explains some of his work in this video (in Spanish).

Cota, R. (2021) *Alejandro Bonifacio Flores*. Hosted by EcoAgTube. 10 minutes.
https://ecoagtube.org/content/alejandro-bonifacio-flores

A version of the section above was previously published on www.agroinsight.com in 2018, by Jeff Bentley.

The Big Mucuna

In 2014, I found myself on the pampas of Argentina. The romantic gauchos were largely gone, although painstakingly reconstructed by a few men who love horses and dapper period costumes. Modern highways and trailer trucks now cross the endless spaces, and the grazing cattle have been largely replaced by large-scale mechanized agriculture: especially of maize and of soy beans, both genetically modified (GM) to resist herbicides.

As explained by two young researchers, Maximiliano 'Maxi' Eize and Patricia Carfagno, in the 1990s, mechanization squeezed some 60,000 smallholders off their land. Most landowners with less than 100 ha no longer farmed their own land, but rented it to pools of investors. The Argentines use the English word 'pool' even when speaking Spanish. *Los pooles* took over the land. Wealthy investors from Buenos Aires gave money to managers who rented small farms and then hired other people to plant them in maize or soy beans, applying some synthetic fertilizer and a lot of herbicides, to save labour. The pool of investors use the land until they have degraded it, and then they give it back to the families that own it, and then they look for more land to spoil, at a profit (Hiba, 2021).

Maxi and Patricia were showing some international extension experts a trial on a farm in La Fe, in Buenos Aires province. The landowners had asked what to do when their farms were trashed and returned.

Maxi and Patricia worked for 5 years in the west of the province, 250 km away, where it was drier. They had successfully tried mixes of vetch, oats and wheat as cover crops, which farmers were then starting to adopt. Now, here in the east of the province, the researchers were starting again; in wetter conditions the results would be different.

The basic idea: after the harvest, the land lies fallow for a few months, building up lots of weeds. GM soy beans are sprayed with lots of herbicides, so only the nastiest species survive. When the land is returned to the owners, it is infested with weeds that are the hardest to control.

The new technology being tried here was to plant wheat, oats, vetch or a mix of vetch and one of the grains. Maxi and Patricia had a whole string of plots with the different treatments. 'Which one works the best?' I asked.

'That's what we'll see at the end of this year', Maxi said, speaking like a true scientist.

But Maxi and Patricia did realize that they needed to come up with a soil conservation strategy that was profitable, even for the greedy pools of investors. Planting vetch and cereals in the off-season must increase yields in the following crop of maize or soy beans, and be cheap to do. Earlier experience on the pampas suggests that an off-season cover crop may increase the maize yield by 1.5 t per hectare, enough to appeal to the cold logic of the pools.

Cover crops are an old idea. Years ago, in Honduras, many farmers tried mucuna, also known as the velvet bean (Buckles *et al.*, 1998). The legume is so robust that when I planted it in my Honduran garden, the mucuna grew right over my orange trees, blanketing them in a thick, green mat.

In the 1990s, I visited Honduran farmers who tried to plant the mucuna in between their maize, only to have to fight back the mucuna as it took over their crop. But farmers and scientists both learn from experience, and they eventually learned to plant mucuna in an off-season, or in an off-year, to give life back to exhausted soils. West African farmers and scientists learned to plant mucuna once the maize is at least 2 months old, so the mucuna benefits from residual soil moisture, but the maize can be harvested before being swallowed up by the mucuna.

I thought that mucuna in some form might work for the land destroyed by the pools. As luck would have it, I just happened to have a DVD in my backpack, with a video on mucuna, filmed with a group of farmers in Togo in West Africa and translated into Spanish. I gave Maxi and Patricia each a copy of the DVD. As experts in cover crops, they had read about mucuna but had never seen it, so they were keen to watch the video.

The pools of investors continue to destroy farmland in Argentina (Hiba, 2021). People who own the land should be the ones who work the land. The land rental 'market' is flawed, and gives perverse incentives to free-riders who mine the soil. Corporate farmers who rent the land, to work it with tractors, motorized herbicide sprayers and GM soy beans, may make efficient use of some workers' time, but they destroy the land, a resource we cannot make more of.*

📹 Related video

Farmers in Togo explain their successful experience with mucuna (as mentioned above).

Agro-Insight (2012) *Reviving Soils with Mucuna*. Hosted by Access Agriculture. 14 minutes. www.accessagriculture.org/reviving-soils-mucuna

---

A version of the section above was previously published on www.agroinsight.com in 2014, by Jeff Bentley.

What Do Earthworms Want?

Even seemingly simple tasks, like raising the humble earthworm, can be done in more ways than one; however, all variations must follow certain basic principles.

In a video from Bangladesh, villagers show the audience how to raise earthworms in cement rings, sunk into the soil. The floor is covered with a sheet of plastic to keep the worms from escaping. The worms are fed on chunks of banana plants, and the ring is covered to keep out the rain but still retain some moisture.

My grandfather used to raise worms in a pressed-board box on his back porch. He fed them strips of newspaper and spent coffee grounds. So, I knew that there was more than one way to raise worms, but I didn't quite realize how many options there were until I saw two small family firms in Cochabamba, Bolivia, this week at an agricultural fair. Both firms raise earthworms and sell the worms, the humus they make and the excess moisture collected in the process (called vermiwash, to use as fertilizer – applied on leaves or the soil).

One company, Biodel, experimented with various types of containers. The worms died in plastic ones, but they thrived inside aluminium cylinders, wrapped in foam (to keep them cool) inside a metal barrel. A screened base with a tray collected the humus, while worm food (especially composted cow manure) was loaded into the top of the barrel.

A second company, Lombriflor, had a different device. They use stacks of plastic-covered wooden trays on a slight slant, and they feed the earthworms residues of maize plants, semi-composted cow manure and kitchen scraps. Earthworms have their favourite foods. 'Earthworms like all of the cucurbits (such as squash), but nothing sour', explained Silvio Gutiérrez, one of the company owners. 'They don't like citrus at all.' Earthworms will eat paper, but they prefer egg cartons.

So here we have a Bangladeshi cement ring, a Bolivian barrel and a set of wooden trays. It seems like a lot of different ways to raise worms, which is an important topic, because the night-crawlers, as my grandfather used to call them, help to decompose the compost and stabilize it, and they improve the soil by indirectly feeding the beneficial microorganisms.

All of these worm brooders share certain core principles. The worms are kept cool, not allowed to escape, fed on organic matter (depending on what is abundant locally) and not allowed to get too dry or too moist.

Earthworms are critical for maintaining soil health and boosting farmers' yields (Fonte *et al.*, 2023). To ensure that more farmers benefit from these beneficial invertebrates, the Bangladeshi earthworm video has been translated into Spanish and 30 other languages.*

Related video

Here is the video on raising earthworms in Bangladesh, mentioned above.

BIID, CARE, DAM and Shushilan (2014) *The Wonder of Earthworms*. Hosted by Access Agriculture. 13 minutes.
www.accessagriculture.org/wonder-earthworms

A version of the section above was previously published on www.agroinsight.com in 2017, by Jeff Bentley.

Reviving Soils

Globally, an estimated 20–60 million hectares of land in developing countries are acquired by foreign companies and investors. This so-called 'land grabbing' has taken place for various reasons, especially the greed for maximum profits. Forests are cut down to produce biofuels, sugar cane, palm oil and soy beans for animal feed. Another reason is to secure food for countries where large areas of land have become unsuitable for farming.

Today, about a third of China's total cultivated area is seriously eroded by wind and water. According to Dave Montgomery in his book *Growing a Revolution,* half of the soil carbon in the Midwestern USA has been lost (Montgomery, 2017). In the European Union, soil erosion affects over 12 million hectares of land – about 7.2% of the total agricultural land – and leads to a 1.25 billion Euros loss in crop productivity (European Commission, 2018).

As people have seen the soil as a warehouse full of chemical elements that could be replenished at will to feed crops, they ignored the microorganisms that help plants to take up the nutrients in organic matter, and soil minerals. Microorganisms do not have chlorophyll to conduct photosynthesis, like plants do, and require organic matter to feed on (Montgomery, 2017).

While acquiring land in other countries as a strategy to secure domestic food supplies has created its own problems, it is hopeful to see that more sustainable initiatives triggered by civil society are gaining momentum, and receiving support from their governments. At various events, President Xi Jinping has declared that China wants to stop destroying natural resources and instead become a global leader for green technologies. By doing so, he has fuelled the aspirations of the Chinese people, over half of whom live in cities, for healthy food. This is reflected in a rapid rise of community supported agriculture (CSA) (Tang *et al.,* 2019).

For several years, the central government of India and various state governments have supported organic and natural farming, with variable degrees of success in scaling (Khurana and Kumar, 2020). Both organic and natural farming rely on biodiversity, on-farm biomass management, natural nutrient recycling, crop rotation, multiple cropping and efficient resource recycling, but they have some differences. Organic farmers may use off-farm purchased organic and biological inputs and natural minerals. Natural farming systems are based on biomass mulching, year-round green cover and on-farm formulations (e.g., made from cow dung and urine, rather than any purchased inputs, even organic or biological ones). By ending the reliance on purchased inputs and loans, natural farming also aims to solve extreme indebtedness and suicides among Indian farmers (Government of India, 2023).

When in 2019 some of our Indian partners produced a farmer training video on how soils can be revived with good microbes, a traditional practice that is now being widely promoted under natural farming, I thought this would be helpful for our garden as well. When we moved into our house in north-eastern Belgium, some of the land had been under intensive cultivation for decades. The soil was hard and had little life. Even though I had mixed some aged cow manure into the planting pits before planting my fruit trees 4 years ago, they struggled during summers that seem to have become dryer and hotter year after year.

I watched the good microbes video from the Access Agriculture video platform and downloaded the factsheet. All I needed was fresh cow dung, cow urine, molasses and chickpea flour. But we don't have cows, only a few sheep, and to have cow dung loaded with good microbes one would have to approach an organic farmer. So, I decided to collect fresh dung from our sheep instead and give it a try.

In the section 'Friendly Germs' (this chapter, below) we write that farmers and farmer trainers in Bolivia mix dung with their hands without any reservations. Likewise, I have often witnessed during my interactions with farmers in South Asia how respectfully they treat dung, as if it were gold. Hence, I started to mix the ingredients. The days before setting up my experiment I had collected my own urine, and because I didn't have molasses to feed the good microbes, I settled for what we had in the house: brown sugar.

Farmers in India also mix leaves of the neem tree into the solution to help control insect pests and diseases. I replaced neem with tansy (*Tanacetum vulgare*), a bitter medicinal plant that we have in our garden. After having added everything to 10 l of water, I placed the drum in the shade, as good microbes don't like direct sunlight.

For 10 days, I let the mixture ferment to increase the number of good microbes, stirring it twice a day to release the gases that could inhibit fermentation. The sweet-sour smell was a good indication that fermentation was successful. The result was a homemade variation of commercially available Effective Microorganisms® and an Indian recipe adapted to Belgian conditions. I kept the filtered solution in recycled plastic milk bottles. Every 2 or 3 weeks I mixed one of the bottles into 100 l of water to then pour the solution around my 30-something fruit trees with a watering can, each tree receiving just enough to moisten the mulch around its base.

Seeing is believing. And doing it yourself adds conviction. In just 6 months the soil around our fruit trees became blacker, softer and crumblier, retaining rainwater much better. I continued the same the year after, and by 2023 had seen that the humus and rich soil life had helped the trees cope much better with the changing climate.

While much of our farmland has been degraded over the past decades, there are solutions to revive our soils. Green technologies spread faster when there is political good will and when farmers have the opportunity to learn from their peers, across borders. Farmer learning videos can help to share these ideas cross-culturally.*

A version of the section above was previously published on www.agroinsight.com in 2020, by Paul Van Mele.

Friendly Germs

At an event in Cochabamba in 2020, just before Bolivia went into lockdown over COVID-19, I had a rare opportunity to see how to make products or inputs for agroecological farming.

The organizers, the NGO *Agroecología y Fe* (Agroecology and Faith), were well prepared. They had written recipes for the organic fertilizers and natural pesticides and provided an expert to explain what each product did and to show the practical steps. The materials for making the inputs were neatly laid out in a grassy meadow. We had plenty of space to build fires, mix materials such as cow dung with earth and water, and to stand and chat. Agronomist Freddy Vargas started by making bokashi, which extensionists have frequently demonstrated in Latin America for decades, especially among environmentally sensitive organizations. Bokashi is sometimes described as fertilizer, but it is more than that; it is also a culture of microorganisms (Quiroz and Céspedes, 2019).

Freddy explained that for the past 25 years, ever since university, he has been making bokashi. He uses it on his own farm, and teaches it to farmers who want to bring their soil back to life. Freddy mixed leaf litter and top soil from around the base of trees (known as *sach'a wanu* – 'tree dung' – in Quechua). The tree dung contains naturally occurring bacteria and fungi that break down organic matter, add life to the soil and help control plant diseases. Freddy added a few packets of bread yeast for good measure. As a growth medium for the microbes, he added rice bran and rice husks, but he said that any organic stuff would work. Next, raw sugar was dissolved in water as food for the microorganisms. He also added minerals: rock flour (ground stone) and *fosfito* (rock flour and bone flour, burned on a slow fire). The pile of ingredients was mixed with a shovel, made into a heap and covered with a plastic tarpaulin to let it ferment. Every day or so it will get hot from fermentation and will have to be turned again. The bokashi will be ready in about 2 weeks, depending on the weather.

It seemed like so much work. Freddy explained that he adds bokashi to the surface of the soil on his farm, and over the years this has helped to improve the soil, to allow it to retain water. 'We used to have to water our apple trees every 2 days, but now we only have to irrigate once a week', he explained.

Next, agronomist Basilio Caspa showed how to make biol, a liquid culture of friendly microbes. He mixed fresh cow dung, raw sugar and water with his hands, in a bucket, a demonstration that perplexes farmers. 'How can an educated man like you mix cow dung with your hands?' But Basilio enjoys making things, and he is soon up to his elbows in the mixture before pouring it into a 200-l barrel, and then filling it the rest of the way with water.

Basilio puts on a tight lid, to keep out the air, and installs a valve he bought for 2 pesos (about US$0.30) at the hardware store. The valve lets out the methane that is released during the fermentation. The biol will be ready in about 4 weeks, to spray on crops as a fertilizer and to discourage disease (as the beneficial microorganisms control the pathogens). Basilio has studied biol closely and wrote his thesis on it. He found that he could mix anything from 0.5 to 2 l of biol into a 20-l backpack sprayer. Higher concentrations worked best, but he always saw benefits whatever the dilution.

We also learned to brew a sulfur-lime mix, an ancient pesticide. This is easy to make: sulfur and lime are simply boiled in water. But do farmers actually use these products?

Then María Omonte, an agronomist with profound field experience, shared a doubt. With help from *Agroecología y Fe*, she had taught farmers in Sik'imira, Cochabamba, to make these inputs, and then helped the communities to try the inputs on their farms. 'In Sik'imira, only one farmer had made bokashi, but many had made biol.' This seasoned group agreed. The farmers tended to accept biol more readily than bokashi, but they were even more interested in the brews that more closely resembled chemicals, such as sulfur-lime, Bordeaux mix (a copper-based fungicide) and ash boiled with soap.

The group excitedly discussed the generally low adoption by farmers of these products. They suggested several reasons: first, the products with microbes are often made incorrectly, with poor results, and so the farmers don't want to make them again. Second, the farmers want immediate results, and when they don't get them, they lose heart and abandon the idea. Besides, making biol and bokashi takes more time than buying agrochemicals, which is discouraging.

Bokashi and biol do improve the soil, otherwise, agronomists like Freddy would not keep using them on their own farms. But perhaps farmers demand inputs that are easier to make and use. The next step is to study which products farmers accept and which ones they reject. Why do they adopt some homemade inputs while resisting others? An agroecological technology, no matter how environmentally sound, still has to respond to users' demands; for example, it must be low cost and easy to use. Formal studies will also help to show the benefits of minerals, microbes and organic matter for the soil's structure and fertility.*

▌▌◀ Related video

Vermiwash, the liquid that drains from compost made by earthworms, is also rich in beneficial microorganisms.

Priya, S. (2020) *Vermiwash: An Organic Tonic for Crops*. Hosted by Access Agriculture. 13 minutes.
www.accessagriculture.org/vermiwash-organic-tonic-crops

A version of the section above was previously published on www.agroinsight.com in 2020, by Jeff Bentley.

Encouraging Microorganisms That Improve the Soil

In 2020, I learned some simple techniques for culturing microorganisms with a few inexpensive ingredients. Ing. Abrahán Mujica showed me and a small group at his agroecology course that you can start by collecting some leaf litter. We gathered the leaves and top soil from the base of two or three molle trees (*Schinus molle*) in the city of Cochabamba.

We put about 5 kg of leaf litter and black soil on a plastic table. We added 1 kg of raw sugar and 1 kg of bran (rich in proteins) to feed the microorganisms, and just enough water to turn the mix to a paste. It should be just moist enough that it will release a couple of drops when you press it in your hand.

As we mixed up the ingredients, a smell like bread yeast soon filled the room.

'Smell the yeast!' Abrahán said. 'The yeast are the first microorganisms to respond to the sugar.'

'Not just yeast', I said. 'There must be 10,000 species of microbes in there.' Abrahán happily agreed.

We filled a third of a 20-l bucket with this paste and covered it with plastic bags, tied on with a rubber tie, to keep out the air. The mix will rot if it is exposed to the air, Abrahán stressed. Fermentation is without oxygen.

After a month, Abrahán mixed the fermented paste with water in a 200-l barrel, sealed it again for another month and then drained off the water, which by then was full of microorganisms.

He filters this solution through an ordinary cloth and bottles the liquid for sale. The label reads 'The Life of the Soil'. It can be sprayed on the soil to make it healthier, or added to compost to speed up decomposition, or used as fertilizer on plant leaves. He said it is intended mainly for soil that has been killed by pesticides, to bring the soil back to life.

Abrahán's home also doubles as a small shop, where he sells *ácido piroleñoso* (liquid smoke distilled during charcoal making – which is mixed with water and sprayed onto crops as natural insect and fungus control). He also makes potassium soap (by mixing potassium sulfate with cooking oil), sulfur-lime blend, Bordeaux mix and other products for protecting plants without toxic chemicals.

Although Abrahán makes the soil-enhancing products he sells, he is happy to teach others. On his agroecology course, he teaches others how to make each product. There will always be lots of people who don't want to make these brews themselves, suggesting that there is a niche for small-scale suppliers who want to make ecological supplies for farmers.*

Related video

Coating seeds with a solution of good microbes gives the plant a healthy start in life.

Pagar, A. and WOTR (2020) *Better Seed for Green Gram*. Hosted by Access Agriculture. 14 minutes.
www.accessagriculture.org/better-seed-green-gram

A version of the section above was previously published on www.agroinsight.com in 2020, by Jeff Bentley.

References

Bonifacio, A., Aroni, G., Villca, M. and Bentley, J.W. (2022) Recovering from quinoa: Regenerative agricultural research in Bolivia. *Journal of Crop Improvement* 37, 687–708.

Buckles, D., Triomphe, B. and Sain, G. (1998) *Cover Crops in Hillside Agriculture: Farmer Innovation with Mucuna*. CIMMYT, IDRC, Mexico DF, Mexico.

European Commission (2018) Soil Erosion Costs European Farmers €1.25 billion a year. Available at: https://joint-research-centre.ec.europa.eu/jrc-news-and-updates/soil-erosion-costs-european-farmers-eu125-billion-year-2018-02-27_en (accessed 16 October 2024).

Fonte, S.J., Hsieh, M. and Mueller, N.D. (2023) Earthworms contribute significantly to global food production. *Nature Communications* 14, 5713.

Government of India (2023) National Mission on Natural Farming Management and Knowledge Portal. Available at: https://naturalfarming.dac.gov.in/ (accessed 16 October 2024).

Hiba, J. (2021) Argentina's land laws stifle sustainable agriculture. *China Dialogue*. Available at: https://chinadialogue.net/en/food/argentinas-land-laws-stifle-sustainable-agriculture/ (accessed 16 October 2024).

Khurana, A. and Kumar, V. (2020) *State of Organic and Natural Farming: Challenges and Possibilities*. Centre for Science and Environment, New Delhi, India.

Montgomery, D.R. (2017) *Growing a Revolution: Bringing Our Soils Back to Life*. Norton, New York.

Quiroz, M. and Céspedes, C. (2019) Bokashi as an amendment and source of nitrogen in sustainable agricultural systems: A review. *Journal of Soil Science and Plant Nutrition* 19, 237–248.

Tang, H., Liu, Y. and Huang, G. (2019) Current status and development strategy for community-supported agriculture (CSA) in China. *Sustainability* 11, 3008.

7 Making the Most of Water

Abstract

Water is becoming an increasingly precious resource, with competing demands for drinking water, agriculture and nature. Agroecological water management in India and Sri Lanka reduces water evaporation through mulching. Drip irrigation in Africa has been limited by project-based approaches, but market demand can be stimulated with farmer-to-farmer learning videos. A case from South Sudan shows that drip irrigation can be successfully adapted using readily available resources. In Jamaica, watermelon farmers leverage pest control services provided by frogs and toads while maximizing the efficiency of water through drip irrigation and mulching.

The Intricacies of Mulching

To make the most of water, one important step is to reduce evaporation by keeping a permanent soil cover, whether by plants or any type of mulch. Everybody working in agriculture knows something about mulching, which can lead us to think that we know all about it. But mulching is a surprisingly complex topic, as I recently realized while backstopping the production of a video from one of the Access Agriculture video partners in India.

For example, different crops may require different types of mulch, and some mulches are better avoided under certain conditions. As with other farming techniques, to make a video on mulch, manuals are often inadequate; one needs to rely on the experience of farmers (Figs 7.1 and 7.2).

We started preparing for the video on mulch during a workshop in Pune, India, in February 2017, where Jeff and I trained several local partners to write factsheets and video scripts for farmers. One of the scripts was on mulch. When I revisit the first draft of that script it is striking how generic our early ideas were.

Among other things, the script mentioned: 'Mulch allows more earthworms and other living things to grow by providing shade. The earthworms make the soil fertile and dig small tunnels that allow the water to go more easily into the soil.' That is all well and good, but that first script was a little

DOI: 10.1079/9781800628793.0007

Fig. 7.1. Farmers with a life-long experience of farming in semi-arid parts of India have deep knowledge of which crop residues to use as mulch.

Fig. 7.2. Dry straw is better than wheat husks to mulch vegetables because husks are easily blown away by wind.

light on how to go about mulching, although it had an idea of using dry straw.

More than a year (and ten versions of the script) later, cameraman Atul Pagar from Pune, India, finished his video *Mulch for a Better Soil and Crop*. For the past few years, with our backstopping, Atul has been producing quality farmer-to-farmer training videos, such as on the use of herbal medicine in animal health. Each of the videos is a testimony to the richness of local knowledge and practices.

For instance, the final version of the video mentions that fruits and vegetables like cauliflower, watermelon and others that grow close to the ground are best mulched with dry straw and sugarcane trash or other crop residues in between every row.

Commonly available wheat husks are not suitable for such crops, as Ravindra Thokal, one of the farmers featuring in the video, explains: 'After harvest, we used to burn the crop residue. Now we do not burn it, but I use it as mulch in my cauliflowers. I do not mulch with wheat husks because they are easily washed away by rain. And when blown away by the wind, the husks can settle on the cauliflowers, which may damage them.'

In less than 12 minutes, the nicely crafted video also explains what to consider when mulching fruit trees, how to fertilize your mulched crop with liquid organic fertilizer, how to control rats that may hide in mulch, and the pitfalls of using plastic mulch. None of these ideas were in the first draft of the video script. The script had been improved over the intervening months by discussing the ideas with farmers and other experts. Although I had read quite a bit about mulching, a lot of the information in the video was new to me.

Farming is intricate. To produce good training videos for farmers requires people who have a keen eye, an open mind and the patience to learn from farmers. Atul has all of these. You can find his videos on the Access Agriculture video platform.*

◼ Related video

Here is Atul's video on mulch.

Pagar, A. and WOTR (2018) *Mulch for a Better Soil and Crop*. Hosted by Access Agriculture. 12 minutes.
www.accessagriculture.org/mulch-better-soil-and-crop

A version of the section above was previously published on www.agroinsight.com in 2018, by Paul Van Mele.

Coconut Coir Dust

Many years ago I wrote one of my first articles, on 'Utilization of coconut coir dust mulch in the tropics', and published it in *Humus News*, a trilingual (Dutch, French, English) magazine from Comité Jean Pain, a Belgian non-profit association that has trained people from across the globe on compost making since 1978 (Van Mele, 1997).

So recently, when one of our Indian video partners decided to make a training video on composting coir dust, I dug up my old article and was pleasantly surprised to see that it still contained a lot of useful information.

Coconut coir dust, or coir pith, is the material that is left over after the fibres have been removed from the coconut husk. Coconut-processing factories often have no idea what to do with this waste, so in many coastal areas in the humid tropics one can find heaps of this natural resource.

Whether economical or ecological motives are the driving force, in low external input agriculture systems in the tropics, farmers often use biowaste for soil conservation and sustainable land use. While coir dust has negligible amounts of nitrogen, phosphorous, calcium and magnesium, making it a poor source of nutrients, it can store up to eight times its dry weight in water. By applying a 15-cm-thick layer of coir dust mulch around coconut seedlings in Sri Lanka, irrigation needs could be reduced by up to 55%. In a pineapple coconut intercrop during the dry season, my coir paper reported that the top soil layer had a moisture content of 49% under the mulch, compared to 10% under a sandy ridge of the same height (Van Mele, 1997).

Weeds in cashew plantations in India are suppressed by applying a layer of 7.5 cm of mulch in a 1.5 m radius around the trees (Kumar *et al.*, 1989). In Sri Lanka, this kind of mulch is mainly used in semi-perennial crops like pineapple and ginger. Coir dust mulch suppressed some of the world's worst weeds, namely goat weed, purple nutsedge and the sensitive mimosa plant (Van Mele *et al.*, 1996).

Besides suppressing weeds, coir dust mulch also helps to establish cover crops. Herbaceous legumes are often used as cover crops under coconut in Sri Lanka, but they are suppressed by weeds in dry weather. Applying coir dust tackles the weeds, but favours the leguminous cover crop during the dry season (Reddy, 2012).

Coir dust consists mainly of lignin, a woody substance which is poorly biodegradable. About 90% is organic matter and the C:N ratio is extremely high, about 100:1 (Prabhu and Thomas, 2002). The low pH of 4.5–5.5 offers an extra protection against biodegradation, as many microorganisms do not survive once the pH drops below 4. Slow biodegradation of organic mulches has recently drawn more attention, especially in the humid and sub-humid tropics, where fast mineralization of the organic matter and leaching of minerals are big problems. While coir dust can easily be applied as a mulch, the recent video suggests that it is better to compost the coir dust first when one wants to use it to improve the soil structure. The video shows how one can easily make one's own organic decomposer from cow dung that is rich in good microbes to break down the lignin. As the good microbes need nitrogen to grow, farmers in the video show how they add decomposed poultry waste. You can also use other material rich in nitrogen, such as urine, Azolla or soy hulls. The techniques shown by farmers resonate well with what researchers like Prabhu and Thomas (2002) have found.

Coir dust is important to control weeds, improve soil physical conditions and retain more water. The dust should be regarded as an important resource for soil conservation and sustainable land use in integrated cropping systems, and not as waste. The use of coir dust in the tropics, however, is not only hindered by a lack of knowledge, which the video aims to share, but is also threatened as coir dust is increasingly exported to Europe where it is used as a horticulture substrate.*

📹 **Related video**

Composted coir pith can be added to the soil, where it degrades slowly and retains five times its weight in water.

Green Adjuvants (2021) *Coir Pith*. Hosted by Access Agriculture. 14 minutes. www.accessagriculture.org/coir-pith

---

A version of the section above was previously published on www.agroinsight.com in 2021, by Paul Van Mele.

To Drip or Not to Drip

Despite huge support, drip irrigation hasn't really taken off in Africa, as I was astonished to learn from Jonas Wanvoeke, who used to work with me at AfricaRice.

I had been convinced of the value of drip irrigation. In 2012 we had made a farmer training video on drip irrigation in Burkina Faso, where Jonas later conducted his field research, and the farmers who featured in the video were all highly convinced of the benefits of drip irrigation and were using it capably.

Jonas's PhD study (Wanvoeke, 2015) showed that the projects all had a pro-poor focus and were often promoting low-pressure kits to irrigate relatively small plots. The technology was always provided by a project without really involving the private sector in Africa. The projects ensured that boreholes (water wells) were installed and that organized farmers worked in groups to manage the kit. Apart from farmers supported by projects like the ones we filmed, Jonas found that hardly anyone used the drip irrigation technology in Burkina Faso, despite two decades of project interventions.

Perhaps the problem with drip irrigation was not the technology itself, but how it had been promoted, as suggested by a recent experience in Benin by Gérard Zoundji.

With the support of Access Agriculture, Gérard embarked on action research; farmer training videos on vegetables were translated into local languages and compiled onto a DVD. One of the videos happened to be on drip irrigation. Gérard sold the DVD at a subsidised rate to agro-dealers and to people who sell entertainment videos. On a note inside the DVD jacket, Gérard printed his phone number so that people could call to ask questions (Zoundji, 2016).

Within 3 months, nearly 400 DVDs had been sold and people were contacting Gérard from as far away as Niger, Nigeria and Ghana. The DVDs were travelling widely, almost under their own power.

On the other hand, when NGOs distribute training videos, almost all of the DVDs stay in the villages that get them.

The farmers who bought the DVDs called Gérard to ask where they could get more videos and where they could get the drip irrigation equipment. Farmers were interested in drip irrigation after all.

As projects gradually come around to the idea that they should work with the private sector instead of avoiding it, they need to understand that handing out gifts to farmers is often counter-productive. The private sector has a role to play, both in selling the hardware and in teaching farmers how to use it (Zoundji, 2016).

Gérard shared with me some photographs from a farmer in coastal Benin who had watched the videos. The farmer had made his own drip irrigation kit using buckets and some old hoses that had served to irrigate the intensive peri-urban vegetable gardens around Cotonou, Benin. Projects promote drip irrigation to save water, but in southern Benin water is plentiful, and available in shallow wells. In this part of Benin, farmers like drip irrigation to save time and labour, not to save water (Zoundji *et al.*, 2018).

So even without the handsome drip irrigation kit presented in the video, farmers see the value of it, and will make the concept work using their own resources. If a video conveys the basic principles that underlie a new technology, the farmers may invent a different way of using the idea.

Projects prefer working with groups of farmers, which only function when the project is there to hold the group together, but drip irrigation is probably best managed by a household.

Development projects offer room for farmers to experiment with novel technologies, like drip irrigation, and working in groups may be the easiest way for the project to get a critical mass of farmers together and to start engaging with the private sector. In a second step, well-made farmer training videos can help to share the innovations with many more creative farm households and to enable the private sector to expand its products and services.*

◼ Related video

The video on drip irrigation discussed in this section can be watched and freely downloaded in more than 40 different languages from the Access Agriculture video platform.

Agro-Insight (2013) *Drip Irrigation for Tomato*. Hosted by Access Agriculture. 12 minutes. www.accessagriculture.org/drip-irrigation-tomato

---

A version of the section above was previously published on www.agroinsight.com in 2016, by Paul Van Mele.

Drip Irrigation Saves Water in South Sudan

In remote areas in post-conflict countries, it may be difficult to get information from universities or extension agencies, but with a smartphone and an internet connection, anyone can watch videos and learn from them. While conducting an online survey of farmers who had previously registered on the Access Agriculture video platform, in 2017 I had a chance to speak on the phone with some highly innovative people, like Isaac Enoch in South Sudan (Bentley *et al.*, 2019).

Isaac Enoch grew up in a village in what was then the south of Sudan, but the worsening war between the north and south drove his family across the border to Uganda. There was little for the kids to do in the refugee camp, so the teenaged Isaac and his friends started to grow vegetables in small patches along the river. When Isaac got enough vegetables to fill a bucket, he would hand the produce to his mother. He told me how impressed he was when she sold the vegetables in the market and came home with money. She began to buy books and shoes for her children, who had been going barefoot. Isaac says this was his first experience of farming as a business.

In 2004, Isaac earned a BSc from Makerere University in Kampala, thanks to scholarships for academic excellence which he was awarded from several UN agencies. He worked for several NGOs in the Sudan until he went on to get an MSc from Bangor University in Wales, UK, in 2007. After graduating, he went straight back to the south of Sudan, and he was there when the new nation of South Sudan was created in 2011, following 20 years of civil war. Isaac was part of a donor-funded project to promote cassava-growing with farmers, but he recalls that the returning refugees were not taking agriculture very seriously. So, he said, 'I'll show them how to do it.' He began growing vegetables on his own, before branching out by giving farmers seed, agreeing on a price once the produce was ready then coming back later to buy the vegetables. During this time, Isaac was working in a rural area, with lots of land, but then violence broke out between different southern ethnic groups and between armed factions that had once been allies in the liberation movement. In these increasingly unsafe conditions, Isaac moved to Juba, the capital of South Sudan.

Land was scarce in Juba, so Isaac started a greenhouse on a small plot. He was not sure how to water his plants. At first, he drew on his own imagination, poking holes in soft drink bottles, filling them with water and placing them near the plants. Then he saw how drip irrigation worked in the video *Drip Irrigation for Tomato* on the Access Agriculture website. He followed the instructions and installed drip irrigation in his greenhouse. In the video, the tanks are filled with hand-carried buckets of water. Isaac was able to fill the tanks with river water, using a small motorized pump.

This worked so well that he also began irrigating some land outside of the greenhouse. He covered the soil with mulch to slow the rate of evaporation and conserve water, an idea he also got from the video.

So much of the food sold in Juba is imported, even the cereals, that anyone who can produce crops locally has a ready market. Isaac is now starting a piggery, producing fodder using hydroponics. He learned about this from a friend, who sent Isaac a link to a video. The original video showed special mechanized trays, but this seemed expensive to Isaac, so he is now growing hydroponic fodder in trays that he designed himself, made by cutting jerry cans in half.

While many projects across Africa have failed to get community groups organized around drip irrigation, access to inspiring training videos can make a difference. Creative, motivated people are able to take ideas from the videos and adapt them to local circumstances.*

Related video

Drip irrigated tomatoes usually need to be staked as well, as shown in this video from India.

Pagar, A. and WOTR (2022) *Staking Tomato Plants*. Hosted by Access Agriculture. 13 minutes.
www.accessagriculture.org/staking-tomato-plants

A version of the section above was previously published on www.agroinsight.com in 2017, by Jeff Bentley.

Toads for Watermelon

The south coast of Jamaica is just right for growing watermelon, where I recently saw the fruit stacked under the shade trees in front of comfortable farm houses. Farmers can earn a tidy living from selling melons on the local market and to the hotels and resorts.

But the trick is to get enough water. In the dry season, a tanker truck will deliver 1000 gallons (almost 4000 l) for US$50. Most of the farmers economize on water by using drip irrigation. For many years, farmers have saved on water by using mulch, made from the lightweight Guinea grass.

Professional crews cut and dry the grass, which is grown in small fields scattered among the patches of watermelon. The grass crews lay out a neat carpet of mulch, which not only keeps the soil moist, but also suppresses weeds and creates a soft, clean bed for the fruit to grow, so it develops an attractive green rind all the way around the fruit. After harvest, the grass decomposes, enriching the soil with organic matter.

I learned about this while visiting Jamaican farmer Junior Dyer with a group of colleagues. We asked when Junior watered his plants. He said at 9 or 10 AM. 'I never water at night', Junior explained, because if he does that, frogs and toads come into the field to eat the insect pests, but then the amphibians stay for the night, digging holes into the moist soil and disturbing the roots. The frogs and toads still come and eat the insect pests when watering is done in the morning, but then they bed down on the edge of the field.

Junior also showed me some of his 13 beehives, which he moves around to pollinate his watermelons, cantaloupe and cucumbers. I asked Junior if he used insecticides to control major insect pests such as whiteflies, thrips and especially aphids, which transmit disease (like watermelon mosaic virus). He admitted, a bit reluctantly, that he did use insecticides. I asked how he managed that without killing his bees. Junior replied that he looks for insecticide labelled as bee-friendly. In truth, most insecticides are toxic for bees. Fortunately, farmers are becoming aware of the dangers of insecticides.

Junior's extension agent, Jermaine Wilson, said that Junior belongs to a farmers' group but that the farmers had already observed on their own that toads and frogs are beneficial creatures. Farmers see them eating insects. Beneficial amphibians are an example of how valuable local knowledge often develops around a topic that is culturally important (like watermelon pests) and easy to observe (like toads eating bugs). I found it encouraging that Junior appreciated the frogs and toads, even though they tend to eat larger insects rather than the really small ones that are the main pests in Jamaican watermelon.

I admired the efficient system the Jamaicans have for producing watermelon, even though they still largely rely on insecticides, with little organic production. But the Jamaican farmers are moving in the right direction by encouraging frogs and toads, and beekeeping will certainly motivate them to further reduce insecticides. Watermelons are a fairly sustainable, commercial crop from family farms. The bees pollinate the melon flowers, and the fruit grows nestled in a bed of mulch, precision-watered with drip irrigation. It's a nice blend of appropriate technology and local knowledge, with frogs and toads contributing along the way.*

◼ Related video

AGRECOL Andes promotes a participatory guarantee system (PGS) to offer farmers a way to organize, receive training and self-guarantee that their quality produce is organic.

Agro-Insight and AGRECOL Andes (2023) *A Participatory Guarantee System*. Hosted by Access Agriculture. 11 minutes.
www.accessagriculture.org/participatory-guarantee-system

A version of the section above was previously published on www.agroinsight.com in 2019, by Jeff Bentley.

References

Bentley, J.W., Mele, P., Barres, N.F., Okry, F. and Wanvoeke, J. (2019) Smallholders download and share videos from the internet to learn about sustainable agriculture. *International Journal of Agricultural Sustainability* 17, 92–107.

Kumar, D.P., Subbarayappa, A., Hiremath, I.G. and Khan, M.M. (1989) Use of coconut coir-pith: A biowaste as soil mulch in cashew plantations. *Cashew* 3(3), 23–24.

Prabhu, S.R. and Thomas, G.V. (2002) Biological conversion of coir pith into a value-added organic resource and its application in Agri-Horticulture: Current status, prospects and perspective. *Journal of Plantation Crops* 30, 1–17.

Reddy, P.P. (2012) *Organic Farming for Sustainable Horticulture*. ISOFAR, Bonn, Germany.

Van Mele, P. (1997) Utilization of coconut coir dust mulch in the tropics. *Humus News* 13, 3–4.

Van Mele, P., Dekens, E. and Gunathileke, H.A.J. (1996) Effect of coir dust mulching on weed incidence in a pineapple intercrop under coconut in Sri Lanka. In: *Proceedings of the 48th International Symposium on Crop Protection — Mededelingen Faculteit Landbouwwetenschappen*, University of Ghent, Belgium, pp. 1175–1180 (Vol. 61).

Wanvoeke, J. (2015) Low-cost drip irrigation in Burkina Faso: Unravelling actors, networks and practices. PhD thesis, Wageningen University, Netherlands.

Zoundji, G.C. (2016) Farmers pay for learning videos. In: Bentley, J.W., Boa, E. and Salm, M. (eds) *A Passion for Video*. CTA, Wageningen, Netherlands, pp. 4–5.

Zoundji, G.C., Okry, F., Vodouhê, D.S. and Bentley, J.W. (2018) Towards sustainable vegetable growing with farmer learning videos in Benin. *International Journal of Agricultural Sustainability* 16(1), 54–63.

8 Ecological Pest and Disease Management

Abstract

In many respects, farmers have a deep knowledge of insect pests and can identify many of them by name, yet farmers are often confused about beneficial insects. This contributes to an over-reliance on pesticides, killing the natural enemies of insect pests and disrupting ecosystems. Agroecological education could encourage farmers to conserve beneficial insects for pest control. For example, innovative farmers use natural techniques like herbal mixtures and ash for pest control, and certain fermented solutions with beneficial microorganisms attract weaver ants which control pests on cashew trees. While researchers can contribute to developing sustainable technologies, there are not enough of them to scientifically validate all of the worthy farmer innovations. Documenting and sharing local innovations can help to promote alternatives to pest management based on toxic chemicals.

Poisoning Our Friends

Except for entomologists, no one knows more about insects than farmers. Wherever researchers have bothered to talk to smallholders about insects, whether in Honduras (Bentley and Rodríguez, 2001), Nepal (Gurung, 2003), with the Dogon in Mali (Griaule, 1961) or the Kayapó of the Brazilian rainforest (Posey, 1984), we see that rural people know the names of hundreds of insects and spiders. This is especially true of organisms that are conspicuous (such as the big ones that are active during the day) or those that make themselves important, e.g. by eating crops (Bentley and Rodríguez, 2001).

However, a recent, quantitative global literature review by Kris Wyckhuys and colleagues (2019) confirms that farmers know little about beneficial insects, especially in industrialized countries. It is fairly easy to notice toads and other relatively large animals eating insect pests (Bentley and Rodríguez, 2001). Many farmers know that birds, frogs and cats are natural enemies of pests. Yet Wyckhuys found that worldwide, farmers mention on average only 0.9 insects or spiders that help to control insect pests.

Farmers can have sophisticated knowledge of certain individual insect species. For example, Van Mele and Cuc (2007) have described Vietnamese farmers who had developed deep knowledge to use weaver ants to control pests in fruit orchards. Such cases are, however, disappointingly rare. Weaver ants are big, diurnal and easy to spot in their treehouse nests sewn together from leaves. Most other natural enemies of insect pests, 'farmers' friends', go unnoticed. Hardly any rural people know about other common natural enemies of pests, such as parasitic wasps, insect-eating fungi and nematodes (see also Van Mele, 2008).

Farmers tend to use more pesticides in cash crops and know fewer natural enemies for these crops than in food staples (Wyckhuys *et al.*, 2019). The use of pesticides is growing worldwide, while the pest problems are as bad as ever (Sharma *et al.*, 2019). Farmers are born experimenters, but to find alternatives to pesticides they need to know more about the natural enemies of insect pests.

Wyckhuys *et al.* (2019) suggest that some of the world's US$0.5 trillion subsidies for agriculture could be devoted to agroecological education. Farmers will never find alternatives to pesticides unless they understand that most insects are beneficial. As farmers use insecticides to kill pests, they unwittingly poison their friends, the insects that eat and kill pests. Some of the subsidies should also be spent on educating consumers, to encourage them to demand pesticide-free food.*

◤◣ Related video

From Africa to the Asia-Pacific, weaver ants are the organic fruit farmers' best friend.

Agro-Insight (2016) *Promoting Weaver Ants in Your Orchard*. Hosted by Access Agriculture. 13 minutes.
www.accessagriculture.org/promoting-weaver-ants-your-orchard

A version of the section above was previously published on www.agroinsight.com in 2019, by Jeff Bentley.

When Ants and Microbes Join Hands

When I recently attended the 1st International Conference on Agroecology – Transforming Agriculture & Food Systems in Africa, one of the research posters on display drew my attention. Effective Microorganisms® are a commercial mix of beneficial bacteria, yeast and other living things. A team in Mozambique had found that the microorganisms not only controlled *Oidium*, or powdery mildew, a serious fungal disease in cashew, but also managed the devastating sap-sucking bug that deforms nuts and causes their premature fall. Or at least, that is what the title said.

Professor Panfilo Tabora had been working for many years with cashew. Not knowing that I was an avid fan of the weaver ant, *Oecophylla*, a tree-dwelling predator, Panfilo gently explained to me that the microorganisms attracted the weaver ant to the cashew trees. 'The ants were a bonus', he said with a smile. I knew that weaver ants effectively control bugs, but now I was completely intrigued: how on earth would microorganisms attract ants?

'Earlier, farmers helped the weaver ants to colonize new trees by putting ropes between trees so the ants could walk to new trees and attack bugs and other pests', Panfilo explained to me. 'But when farmers started spraying fungicides, the ants disappeared.'

For several years, Panfilo and his colleagues began to teach villagers to make their own liquid molasses from dried and stored cashew apples as a source of sugar, minerals and amino acids to feed and multiply the microorganisms. So, the farmers made molasses to feed the Effective Microorganisms®, which controlled the *Oidium*. But even when the fermented solution was ready to spray on the trees it was still sweet. 'When farmers spray their trees with the solution, the sweet liquid and amino acids attract the ants.'

Although the poster did not tell the full story, there was still truth in saying that microorganisms controlled the fungal disease and the pest; in reality it was the fermented solution that attracted the ants, which controlled the bugs. Still, even such a roundabout pest control is worth having.

I felt reassured to know that valuable ancient technologies of biological control, such as weaver ant husbandry, have a future when combined with modern agroecological technologies that restore rather than kill ecosystems.

'And we discovered a few more unintended benefits', Panfilo continued. 'By spraying the tree canopies with microorganisms, farmers are no longer exposed to pesticides and can reduce the cost of pruning.' As pesticides are expensive and harmful, farmers need to move quickly from one tree to the next to spray the outside canopy of the trees, or else they will get covered with chemicals. But as Effective Microorganisms® are safe for people, farmers can actually spray the undercanopies from below. The tree canopies often touch one another, which also helps the ants to move between trees. Instead of pruning every year, Panfilo's team tells farmers to just prune once every other year, or even every 3 years, so as to have more terminals for flowering and fruiting and to let the ants move from tree to tree. All of this adds up to more yield.

At that stage, I was so impressed that I had a hard time absorbing yet another unintended benefit of this organic technology. In Mozambique, as in many other countries, farmers use the fallen cashew apples to make cashew apple juice. 'By spraying cashew trees with Effective Microorganisms®, it acts as an antioxidant so the juice retains its clear colour for at least 2 months', said Panfilo.

Quite a few of the presentations at the conference had nicely illustrated the benefits of organic agriculture to people and the environment, but Professor Panfilo Tabora and his team stood out because they illustrated how the introduction of even a single modern ecotechnology can have such a wide range of benefits.

Not all microorganisms are bad, as people in the industry, schools and media often want to make us believe. Thanks to the work of practical researchers, we learn that this healthy mix of microscopic flora can cure mildew, attract ants that kill pests, provide a safe alternative to pesticides and stop cashew fruit juice from oxidizing for months.*

📹 **Related video**

Fruit farmers in West Africa control fruit flies with weaver ants and other natural techniques.

Agro-Insight (2016) *Weaver Ants against Fruit Flies*. Hosted by Access Agriculture. 11 minutes.
www.accessagriculture.org/weaver-ants-against-fruit-flies

A version of the section above was previously published on www.agroinsight.com in 2019, by Paul Van Mele.

Innovating with Local Knowledge

Local knowledge is dynamic and farmers are fast to adapt traditional practices when the need arises, as we saw during a recent filming visit.

The fall armyworm (*Spodoptera frugiperda*) arrived in Africa only in 2016 and is creating panic among farmers and governments alike. International development organizations are quick to call for public funds to respond to evident emergencies.

But farmers can't always wait for solutions to be developed by researchers or for government support. John Fundi from Embu County, Kenya, combined various observations on how ants behave to develop his own solution. For example, ants like fat and caterpillars, so if you smear fat on the maize stalks you can attract the ants to move up onto the plants and eat the caterpillars.

Farmer Aaron Njagi shared another interesting innovation based on keen observations. As an herbalist, Aaron knows a lot about which plants can be used to cure people and which ones can be used to kill or repel insect pests. The herbal pesticide that he makes and uses to kill caterpillars in his vegetable crops proved inadequate to control the fall armyworm, so Aaron immediately figured that this pest was not like any other. His herbal mix needed extra strength.

'Just one drop of aloe vera in water is enough to cure people from respiratory problems, so I decided to add the strength of this plant to the mix of plants I use to control the other caterpillars', he says. On top of that, he adds chopped chilli for extra bitterness and strength, and then boils the lot. Once the water has cooled down a little, Aaron removes the plants from the water and adds a little snuff tobacco.

'After fermenting the mix for a week in the shade, I can now use it', he continues, 'but you need to dilute it as it is very powerful. I also decided to add a little washing powder before spraying it, so it sticks better to the maize plants.'

Farmers know when something works and when something doesn't work. Everywhere we went, we heard that pesticides did not kill the fall armyworm. But Aaron's mixture works. That he is already asked by his neighbours to spray their fields with his herbal medicine further testifies to how fast farmers can innovate.*

▐◀ Related video

One of two videos on fall armyworm, developed in collaboration with the Food and Agriculture Organization of the United Nations (FAO) with funding from the McKnight Foundation's Collaborative Crop Research Program (CCRP).

Agro-Insight and FAO (2018) *Killing Fall Armyworms Naturally*. Hosted by Access Agriculture. 16 minutes.
www.accessagriculture.org/killing-fall-armyworms-naturally

A version of the section above was previously published on www.agroinsight.com in 2018, by Paul Van Mele.

Ashes to Aphids

Anyone interested in organic farming will eventually come across the use of ash to protect crops from pests and diseases (e.g. Stoll, 1988). The internet has made it easy for people to consult and to copy each other's training materials. But one has to be cautious when borrowing ideas, as we recently learned during a scriptwriting workshop in Bangladesh.

During the first day of the course, the 13 trainees from Bangladesh and Nepal laid out their key ideas to write a factsheet and a script on a particular problem. All of our script ideas were hot topics, problems that occur widely across developing countries, requiring good training materials with ideas that are both feasible for smallholders and environmentally friendly.

One of the selected topics was how to manage shoot and fruit borer in aubergine (eggplant), a pest for which many farmers in South Asia spray pesticides twice a week or more. Just knowing this makes you frown when this tasty vegetable is presented to you in one of the delightful dishes when visiting South Asia.

Another group worked on aphids in vegetables and suggested using ash to manage these pervasive pests. When Jeff and I asked why ash is useful, the group gave us various reasons: because it is acidic; it contains sulfur; it is a poison; the ash creates a physical barrier which prevents the aphids from sucking the sap of the plant. These all sound like plausible answers, yet some are incorrect. Ash is rich in calcium, like lime, and therefore not acidic, for example (Gill *et al.*, 2015).

The cuticle that covers the bodies of insects generally has a wax coating, to prevent dehydration. Depending on the particle size, shape and pH, wood ash can abrade the insect's wax, and even the rest of the cuticle, causing death by desiccation. Ash also sometimes serves as a physical barrier to the insects' movement (Vincent, 2001; Batistič *et al.*, 2023).

The FAO's website on applied technologies (TECA – Technologies and Practices for Small Agricultural Producers) suggests controlling aphids by applying wood ash after plants are watered. If not, the sun may cause the leaves to burn. Our simple question about using ash reminded me that the scientific basis for many local innovations is poorly understood. There are too few researchers to validate each technology, and limited resources often focus on high-tech solutions (e.g. plant breeding) rather than low-tech farmer innovations.

We may not always know why local innovations work, which is all the more reason to be cautious when recommending substitutions. During this workshop, for instance, I learned that not all ashes are the same. Shamiran Biswas, an extensionist with a rich experience of working with farmers across the country, explained: 'When one field officer told farmers to sprinkle ash on his crop, a farmer who followed this advice saw his entire bean field destroyed within half an hour. We were shocked and tried to figure out what went wrong. It seemed that the farmer had used ash from mustard leaves, which some rural women add to their cooking fires when they are short of wood. But leaf ash from mango, mustard, bamboo and other plants may also be harmful when sprinkled on crops. The only ash other than from wood that is fully safe to recommend is ash from rice straw or rice bran', Shamiran concluded. He added that 'the ash should be cold and sprinkled on the crop when the leaves are still wet from the morning dew.'

Experienced extension agents like Shamiran are experts at explaining farmers' ideas to outsiders, as well as explaining scientific ideas to rural people.

When people give advice to farmers, or develop farmer training materials, it is easy to copy ideas from the internet. It is easy to assume that because ash is natural that it must be harmless.

A natural solution can go wrong, even one as simple as applying ash. To develop good farmer training videos, solid interaction with farmers is crucial. And collaboration with a seasoned, open-minded extensionist helps to orient us in the right direction.*

🎥 Related video

After finishing the course discussed in this section, some of the participants filmed the following video, now available in 28 languages.

CCDB (2019) *Managing Aphids in Beans and Vegetables*. Hosted by Access Agriculture. 9 minutes.
www.accessagriculture.org/managing-aphids-beans-and-vegetables

---

A version of the section above was previously published on www.agroinsight.com in 2017, by Paul Van Mele.

Killing Mealybugs with Bananas

Most scientists work in disciplinary fields, a narrow focus that encourages researchers to promote what they believe in the most and discard alternatives. Thailand has made headway in controlling the cassava mealybug, a pest that arrived in the country in 2008. While introducing a parasitic wasp contributed to controlling the cassava mealybug, it was only part of the solution, as we learned on a recent visit to make a farmer training video. Staff from the Department of Agricultural Extension embrace classical biological control, such as introducing parasitic wasps, but seemed sceptical about farmers using botanical pesticides.

Farmer innovations are often discarded because they have not been proven scientifically. But scientists may never have the interest, time or resources to validate all farmer innovations. And scientific validation is not always needed before one can promote a local innovation. Most of the other practical arts are not subject to scientific validation either.

Mr Sawart Jaimetta lives in the small village of Nonemakharpom near Khon Buri in north-eastern Thailand. He earns a living from growing cassava and rearing crickets, and grows a variety of plants around his house for food, medicine and spice. Like many farmers in developing countries, Mr Sawart creatively uses all available resources.

Before planting his cassava cuttings, Mr Sawart soaks them in a solution of water and an extract made from banana plants. According to him, it kills all the mealybugs hiding in the cassava buds and increases the vigour of the cuttings.

'There are beneficial microorganisms at the base of the banana stems. When we want to make a plant extract we have to dig the banana stems in the early morning. When the plant has not yet received sunlight, the hormones are still at the base of the stem', Mr Sawart says, his choice of vocabulary suggesting how he has creatively blended outside knowledge with his own keen know-how (Fig. 8.1).

Mr Sawart chops the corms and bottom halves of two young banana shoots into small pieces. He mixes 10 l of molasses with 10 l of water in a bucket to which he adds a small bag of Effective Microorganisms® to speed up the decomposition. He says the mix would also work without adding these beneficial bacteria, but it would take longer. If this seems far-fetched, traditional wine and sourdough bread are made with wild microorganisms.

After stirring this solution, Mr Sawart pours it onto the banana cuttings in a plastic drum. The drum is tightly closed and placed in a shady place. Every week, he stirs the solution to speed up decomposition. A few months later, a white film covers the surface and the extract is ready for use.

As the extract is powerful, Mr Sawart mixes one litre of it with 200 l of water before drenching the cassava cuttings for 10 minutes (Fig. 8.2).

Fig. 8.1. Beneficial microorganisms and hormones are abundant at the base of the banana stems.

'If we use banana shoot extract to soak cassava cuttings, they will sprout twice as fast, in 5 days. We can also use the extract to spray the crop's leaves. It is like a hormone that makes the cassava grow well and strong to resist mealybugs', says Mr Sawart.

Mr Sawart received training from various projects and added his own experience: 'I have also made extracts from other plants, but the corms and base of banana stems give the best results.' To kill the cassava mealybug, he also rears *Beauveria* (a fungus that kills insects), parasitic wasps (that lay their eggs inside insect pests) and green lacewings (Chrysopidae), whose larvae eat mealybugs and other pests.

As this story shows, farmers are not restrained by scientific disciplines. Farmers need training and new ideas to test. They will apply whatever does the job, especially if it is low cost and of little risk to their health. When we communicate with farmers, especially through mass media, we need to open up to solutions offered by different disciplines as well as to those developed by farmers.*

Fig. 8.2. Mr Sawart Jaimetta in Thailand soaks cassava cuttings in a fermented solution of microorganisms to control mealybugs.

🎥 **Related video**

Farmers in Bangladesh show how to culture good microorganisms to improve their crops.

Siddique, R.K. (2022) *Healthier Crops with Good Microorganisms*. Hosted by Access Agriculture. 16 minutes.
www.accessagriculture.org/healthier-crops-good-micro-organisms

---

A version of the section above was previously published on www.agroinsight.com in 2014, by Paul Van Mele.

Good Fungus for Healthy Groundnuts

Diseases need to be cured; this is true for people, animals and plants. In plant protection, fungicides are probably more readily seen as acceptable than insecticides, which are well known to harm the ecosystem, bees, birds and people. But plants can be protected without chemicals, as people from the

M.S. Swaminathan Research Foundation (MSSRF) in India are showing in their steadily growing series of farmer training videos.

Their latest farmer training video on root and stem rot in groundnut nicely shows how beneficial fungi like *Trichoderma* can control root and stem rot diseases without the need for chemical fungicides. Indian farmer Govindammal shows the viewer how she carefully coats the groundnut seed with *Trichoderma*, using some water to make the powder stick to the seed. She mixes it on a jute bag without using her hands, to avoid breaking the seed.

Some farmers add *Trichoderma* directly to the soil by mixing it in the manure. For 1 ha of land, they mix 2 kg of *Trichoderma* with ten baskets of farmyard manure. They leave the mix for a day in the shade before applying it to the field. The good fungi will grow faster with the manure. By broadcasting this mix on their field before sowing, farmers will grow abundant, healthy groundnuts.

Biological pest control was long restricted to insects, so when doing a Google Scholar search on root and stem rot in groundnut, I was pleasantly surprised to see that many top articles are on biological control with beneficial fungi such as *Trichoderma*. Indian scientists have dominated this research and hence it comes as no surprise that in India *Trichoderma* has become widely available as a commercial product.

Apart from their own videos, MSSRF staff have also translated farmer-to-farmer training videos that were produced in Bangladesh and Africa. MSSRF makes the Tamil versions of the videos available to farmers through its rural plant clinics and farmer learning centres.

Extension agents can and do make a difference in farmers' attitudes towards agrochemicals, but videos can speed up this process. Besides, quality training videos will change the behaviour not only of farmers, but also of extension staff and some researchers.

Hopefully, in the future, we will see more research and extension in support of organic and ecological agriculture to help more farmers wean themselves away from chemical-based production of food and fibre. As we have seen with other technologies such as drip irrigation, farmer training videos can create a real demand for green technologies and trigger rural entrepreneurs to invest in them.*

▶ Related video

This video by MSSRF shows how to manage mealybugs without agrochemicals.

MSSRF (2019) *Managing Mealybugs in Vegetables.* Hosted by Access Agriculture. 11 minutes.
www.accessagriculture.org/managing-mealybugs-vegetables

A version of the section above was previously published on www.agroinsight.com in 2019, by Paul Van Mele.

Bullets and Birds

Chatting with Vera Kuijpers from the organic farm *Het Eikelenhof* in the north-east of Belgium, I was reminded again that not only have we forgotten the importance of many crops, but often farmers struggle with forgotten pests that have been neglected by formal research.

Vera took me on a tour of the farm. A line of 6-year-old blossoming apple trees marked the edge of the farm, while a nature reserve of woodlots and heath marked the far end. Each apple tree had a wooden pole next to its stem, with a rusted tin can on top of it.

'This is just like farmers in Africa do to protect their wood from infiltrating rain', I said. By protecting the tip of the pole, the wood wouldn't rot as fast.

'No,' said Vera, 'we use this to scare away mole rats who chew the roots of the apple trees. The mole rats don't like the vibrations made by the empty cans in the wind, but it is not very effective. At times my husband Johan can easily lift up an apple tree. All the roots are gone.'

It reminded me of a trip I had made to Uganda just a few weeks earlier, where bean farmers mentioned birds, rats and moles as their priority pests. So, I asked Vera what other major pests she faced on her farm.

'Unlike other farms in this area, the seed we use has not been treated with chemicals. So, when the cabbage seedlings come up, we have a big problem with wood pigeons. We then call some of the local hunters who come and diagnose the damage. If they see it is pigeon damage, they will come and guard the crop in the early morning. When a flock of pigeons arrives, they will shoot in the air to scare away the birds. And if the birds have the courage to return one more time, they will shoot a pigeon. After that, that flock will never come back again.'

The owners of the organic farm are good friends with all people in the community, and at times can rely on such needed services as the early morning hunters, who are sensitive enough to shoot just one bird.

Farmers across the world develop creative solutions to manage pests. Documenting these has been one of the passions of the Agro-Insight team.

While we train local partners to develop high-quality videos with and for farmers, a large part of our efforts focuses on teaching our trainees to listen to farmers, who have an integrated view of the farm, and often have excellent ideas, for example about controlling birds, rodents and other vertebrate pests. An earlier video made by the Beninese NGO the Organization for Sustainable Development, Strengthening and Self-Promotion of Community Structures (DEDRAS – *Organisation pour le Développement Durable, le Renforcement et l'Autopromotion des Structures Communautaires*) on soy tells a similar story of farmers engaging local hunters to protect their crop from rabbits.*

📹 **Related video**

The video made by DEDRAS can be seen on the Access Agriculture website.

DEDRAS (2016) *Soya Sowing Density*. Hosted by Access Agriculture. 11 minutes.
www.accessagriculture.org/soya-sowing-density

---

*A version of the section above was previously published on www.agroinsight.com in 2015, by Paul Van Mele.

You Can't Poison Your Weeds and Eat Them Too

'What about the edible weeds?' I asked, near the end of a pleasant day, chatting with a group of friendly, articulate smallholders near Retalhuleu in lowland Guatemala. Keith Andrews and I were doing a study of local and scientific knowledge (Bentley and Andrews, 2011). The farmers had collected wild plants from the area, and taught us about each one. But I knew from the literature that many wild plants in Guatemala and southern Mexico are *quelites* or edible weeds, nutritious, leafy vegetables that grow wild in the maize field.

Then the people explained with a touch of sadness how they had completely lost the wild edible greens that once grew abundantly. 'When we started to use herbicides, the quelites disappeared completely.'

At four sites around Guatemala, wherever people had adopted herbicides, they had selected for a tough, robust community of weeds that were hard to kill and good for nothing.

The community of Palestina, near the wondrous Lake Panajachel, was one of the places that did not use herbicides. As the local people walked with us around their fields and gardens, they spontaneously picked wild plants to take home. Most of the plants had uses.

And while ethnobotanists often like to write about medicinal plants and leafy vegetables, by far the most common use of edible weeds is as livestock fodder (Bentley *et al.*, 2005). These weeds for feed come at the most crucial time (for people and their animals), before the harvest, when the stored food has been eaten up and the new crop is not quite ripe yet. Edible weeds supplement the food of people, but they also tide the animals over the hungry season.

Hand weeding a maize field with a hoe is tedious and backbreaking, and many farmers have come to perceive herbicides as a magic wand. The farmer straps on the backpack sprayer, pumps it up and waves the wand over the weeds, while the herbicide solution springs like a mist from the nozzle on the end, and makes all the hard work vanish.

However, the chemicals exterminate the delicate weeds, the ones that people and animals can eat. This opens up a niche for the most notorious, herbicide-resistant weeds that nature has in stock. Elsewhere in this book, we mention innovative farmers who use alternative methods to control weeds, such as intercropping and cover crops (see especially Chapters 1 and 6, this volume).*

◼️ Related video

In this video, farmers in northern Nigeria show how to reduce damage by the parasitic weed striga by intercropping grains with legumes.

Agro-Insight, CBARDP, ICRISAT and KNARDA (2016) *Growing Row by Row*. Hosted by Access Agriculture. 9 minutes.
www.accessagriculture.org/grow-row-row

---

*A version of the section above was previously published on www.agroinsight.com in 2014, by Jeff Bentley.

References

Batistič, L., Bohinc, T., Horvat, A., Košir, I.J. and Trdan, S. (2023) Laboratory investigation of five inert dusts of local origin as insecticides against the Colorado potato beetle (*Leptinotarsa decemlineata* [Say]). *Agronomy* 13, 1165.

Bentley, J.W. and Rodríguez, G. (2001) Honduran folk entomology. *Current Anthropology* 42, 285–301.

Bentley, J.W. and Andrews, K.L. (2011) *Los Dos Saberes. La Sinergia entre los Saberes Científicos y Locales: Un Diálogo entre Técnicos Agropecuarios y Productores para Mejorar la Extensión e Investigación en Guatemala (Two Ways of Knowing. The Synergy Between Scientific and Local Knowledge: A Dialog Between Farmers and Agricultural and Livestock Technical People to Improve Research and Extension in Guatemala)*. IICA and CIDA, Guatemala City, Guatemala.

Bentley, J.W., Webb, M., Nina, S. and Pérez, S. (2005) Even useful weeds are pests: Ethnobotany in the Bolivian Andes. *International Journal of Pest Management* 51, 189–207.

Gill, K.S., Malhi, S.S. and Lupwayi, N.Z. (2015) Wood ash improved soil properties and crop yield for nine years and saved fertilizer. *Journal of Agricultural Science* 7, 72–83.

Griaule, M. (1961) Classification des insectes chez les Dogon (Classification of insects among the Dogon). *Journal des Africanistes* 31, 7–71.

Gurung, A.B. (2003) Insects – a mistake in God's creation? Tharu farmers' perception and knowledge of insects: A case study of Gobardiha Village Development Committee, Dang-Deukhuri, Nepal. *Agriculture and Human Values* 20, 337–370.

Posey, D.A. (1984) Hierarchy and utility in a folk biological taxonomic system: Patterns in classification of arthropods by the Kayapó Indians of Brazil. *Journal of Ethnobiology* 4, 123–139.

Sharma, A., Kumar, V., Shahzad, B., Tanveer, M., Sidhu, G.P.S. *et al.* (2019) Worldwide pesticide usage and its impacts on ecosystem. *SN Applied Sciences* 1, 1–16.

Stoll, G. (1988) *Natural Crop Protection Based on Local Farm Resources in the Tropics and Subtropics*, 3rd edn. Margraf Publishers Scientific Books, Weikersheim, Germany.

Van Mele, P. (2008) The importance of ecological and socio-technological literacy in R&D priority setting: The case of a fruit innovation system in Guinea, West Africa. *International Journal of Agricultural Sustainability* 6, 183–194.

Van Mele, P. and Cuc, N.T.T. (2007) *Ants as Friends: Improving your Tree Crops with Weaver Ants*. CAB International, Wallingford, UK.

Vincent, J.F.V. (2001) Cuticle. In: Buschow, K.H.J., Cahn, R.W., Flemings, M.C., Ilschner, B., Kramer, E.J. *et al.* (eds) *Encyclopedia of Materials*. Science and Technology. Pergamon, Oxford, UK, pp. 1924–1928.

Wyckhuys, K.A.G., Heong, K.L., Sanchez-Bayo, F., Bianchi, F.J.J.A., Lundgren, J.G. *et al.* (2019) Ecological illiteracy can deepen farmers' pesticide dependency. *Environmental Research Letters* 14, 093004.

9 Healthy Livestock

Abstract

Keeping livestock healthy is a crucial challenge for family farms everywhere. For example, in the Netherlands intensive farming has led to environmental degradation and reliance on veterinary drugs. Dutch veterinarian Katrien van't Hooft and colleagues are working with farmers to introduce 'natural livestock farming' (NLF) as a solution. In India, using a transdisciplinary approach, veterinarians are also learning to combine traditional practices with scientific approaches. In Nigeria, Fulani herders once used herbal remedies for livestock ailments, but younger generations have largely abandoned these practices in favour of veterinary drugs, raising concerns about antibiotic resistance and the loss of valuable knowledge. In Bolivia, the *Yapuchiris* are expert farmers, who actively share and validate local knowledge, like herbal salves for wounds, ensuring the preservation and wider application of traditional veterinary practices. Peri-urban Ugandan farmers use good microorganisms to keep animal bedding healthy and odour-free.

Against or with Nature

Ask any tourist what comes to mind when they think of the Netherlands and many will say 'windmills'. Ask any agricultural professional what the Netherlands is known for and they may mention 'water management' and 'dairy' (you know, the big round cheeses, Fig. 9.1). Few people realize how these are all intricately interwoven, and how their interaction over time has created an environmental disaster that is only now being painstakingly addressed.

In his thought-provoking book *Against the Grain*, James Scott (2017) draws on earlier work of anthropologists and archaeologists to shed light on how early humans changed their environment to get more food from places closer to home. The very act of domesticating plants, animals and fire in a sense also domesticated us as a species. While modern cows and many of our crops can no longer survive without us, we can no longer survive without them. Besides fire, people are also heavily dependent on water. In fact, everywhere in the world,

Fig. 9.1. Cheese makers and consumers in the Netherlands benefit from natural animal health care by dairy farmers. Photograph used with permission from Katrien van't Hooft, Dutch Farm Experience.

ancient peoples first settled near rivers or at the fringes of wetlands which, along with the nearby forests, provided a rich variety of food (Fagan, 1996).

Agricultural technology was fairly stable for centuries, but slowly began to change in medieval times, which brings us back to the windmill. Fixed windmills to grind grain were found in Flanders by the 11th century. In the 16th century, following further technical innovations, windmills could also be used to saw wood and to pump water (McNeil, 1990). Soon the Dutch landscape was dotted with thousands of windmills. The now so typical landscape of peat grasslands and ditches is a manmade ecosystem shaped through drainage by windmills. The new pastures with lowered groundwater tables were especially apt for dairy farming, serving what became the world-renowned Dutch dairy sector.

The drainage of the wetlands sounds like a great agronomic achievement, but a Dutch veterinarian, Katrien van't Hooft, director of Dutch Farm Experience, recently showed me the other side of the coin. The continuous drainage of surface water and lowered groundwater table, combined with modern dairy farming and use of tractors, has caused a drop in the peatland. The land has been sinking several centimetres per year for a long time, faster than the rise in sea level. Projections are that under current management the peat soils will further sink 2 m before 2050 and become a major threat to the country. Although the Dutch government is taking urgent measures to restore the groundwater table, the challenges do not stop there (Dutch Farm Experience, 2021).

As drained peat releases carbon dioxide (CO_2), the Dutch government has set up a scheme to reward farmers who help raise the groundwater table. But wet pastures require a very different management, as farmers are now beginning to learn. When collecting hay on wet pasture, overloaded machines risk getting stuck. Maize cannot be grown, because this water-loving crop lowers the groundwater level in the peat land. The black-and-white Holstein-Friesian cow, commonly used in the Netherlands for its high milk production, requires maize and concentrated feed. In the peat lands it is therefore now being crossed with 'old-fashioned' local cattle breeds, such as Blister Head (Blaarkop) and MRY (Maas-Rijn-Ijssel breed) (Katrien van't Hooft, personal communication, Netherlands). These so-called dual-purpose cows yield milk and meat, perform well on plant-rich pastures and have the benefit that they can produce milk with minimal use of concentrated feed.

However, as the peat pastures need to become wetter again, these cows are increasingly suffering from some 'old diseases', including intestinal worms and the liver fluke, which spends part of its life cycle in mud snails. Farmers are using anthelmintics (anti-worm chemicals) to control this, but the anthelmintics to control liver fluke are forbidden in adult cows, for milk safety reasons. Moreover, just as with antibiotics, the internal parasites are quickly building up resistance against anthelmintics, and the dairy sector is being forced to rethink its position of always trying to control nature (Katrien van't Hooft, personal communication, Netherlands).

Now here comes a twist in the story. As Katrien explained to me, these common animal diseases used to be managed by appropriate grassland management, use of resilient cattle breeds and strategic use of (herbal) medicines. But most of this traditional knowledge has been lost over the past decades.

With a group of passionate veterinary doctors and dairy farmers, Katrien has established a network with colleagues in the Netherlands, Ethiopia, Uganda and India to promote natural livestock farming, or NLF (Natural Livestock Farming Foundation, 2021).

Inspired by ethnoveterinary doctors from India, Dutch veterinary doctors and dairy farmers have gained an interest in looking at herbs, both for animal medicine and for enriching grassland pastures to boost the animals' immune system (Fig. 9.2). Together they have developed the so-called NLF 5-layer approach to reduce the use of antibiotics, anthelmintics and other chemicals in dairy farming.

Resistance to chemical drugs used in livestock, whether against bacteria, fungi, ticks or intestinal worms, will have a dramatic effect on people. For example, the bacteria that gain resistance to antibiotics in animals become 'superbugs' that are also resistant to antibiotics in human patients. The abuse of antibiotics in livestock can ruin these life-saving drugs for people.

Scott (2017) describes in his book that when we started intensifying our food production thousands of years ago, we lost an encyclopaedia of knowledge

Fig. 9.2. Diverse pastures rich in herbs are essential for healthy cows. Photograph used with permission from Katrien van't Hooft, Dutch Farm Experience.

based on living with and from nature. In the same vein, traditional knowledge of agriculture has been eroding since the mid-20th century, with intensification brought on by machinery and chemicals, like the Dutch dairy farmers who lost most of their folk knowledge about plants and the 'old' cattle diseases.

While the challenges are rising, it is fortunate that 21st-century humans are able to learn from each other's experiences at a scale and speed unseen in history. Dutch dairy farmers are not the only ones to have lost traditional knowledge. It has happened across the globe, and more efforts are needed to help make such worthwhile initiatives of knowledge-sharing go viral (in a manner of speaking).*

📹 **Related video**

For centuries, farmers in India have known about and used herbal medicines for livestock health.

Pagar, A. and Anthra (2023) *Herbal Medicine against Foot Rot in Livestock*. Hosted by Access Agriculture. 12 minutes.
www.accessagriculture.org/herbal-medicine-against-foot-rot-livestock

A version of the section above was previously published on www.agroinsight.com in 2021, by Paul Van Mele.

Veterinarians and Traditional Animal Health Care

In Pune, Maharastra, the Indian NGO Anthra has devoted much of its energy to documenting traditional animal health knowledge and practices across India. Dr Nitya Ghotge, along with a team of women veterinarians, founded Anthra in 1992 to address the problems faced by communities that rear animals, particularly peasants, pastoralists, adivasis (indigenous peoples of South Asia) and dalits (formerly known as untouchables, who are outside the caste system). The women veterinarians of Anthra continue to blend local and scientific knowledge, working with women and others who remained hidden from the gaze of mainstream development (Anthra, 2023).

In their encyclopaedia *Plants Used in Animal Care*, Anthra has compiled an impressive list of plants used for veterinary purposes and fodder (Ghotge and Ramdas, 2008).

To ensure that local communities across the Global South benefit from this indigenous knowledge, Anthra started collaborating with one of Access Agriculture's trained video partners (Atul Pagar) to gradually develop a series of farmer-to-farmer training videos on herbal medicines.

While Indian cities are booming and the agro-industry continues its efforts to conquer lucrative markets, many farmers and farmer organizations across the country treasure India's rich cultural and agricultural heritage. Unfortunately, this is not the case everywhere. In many countries, local knowledge is quickly eroding as the older generation of farmers and pastoralists disappear.

A few years ago, I was thrilled to work with traditional Fulani herders in Nigeria, only to discover that none of them still used herbal medicines. Even to treat something as simple as ticks, the young herders confidently turned to veterinary drugs. Although the elder people could still readily name the various plants they used to treat various common animal diseases, the accessibility and ease of use of modern drugs meant that none of the herders applied herbal medicines anymore. The risks of such drastic changes quickly became apparent. As we were making a series of training videos on quality milk, which should have no antibiotics or drug residues, we visited a hospital to interview a local doctor.

'If people are well, they are not supposed to take antibiotics. If such a person is sick in the future and the sickness requires the use of antibiotics, it would be difficult to cure because such drugs will not work. It can even make the illness more severe', doctor Periola Amidu Akintayo from the local hospital confided in front of the camera.

Later on, we visited a Fulani cattle market. For years, these markets have been bustling places where the semi-nomadic herders meet buyers from towns. People exchange news on latest events and the weather, but above all assess the quality of the animals and negotiate prices. Animals that look unhealthy or have signs of parasites obviously fetch a lower price. Given that the cattle market is where the Fulani herders meet their fellow herders and clients, I quickly realized why the entire market was surrounded by small agro-vet shops. Competition was fierce, and demand for animal drugs was high.

Modern drugs come with an enclosed instruction sheet, but as with pesticides, nobody in developing countries reads this advice. To keep costs down, many herders and farmers administer drugs to their own animals without consulting a veterinarian. Perhaps even more worrying, few people are aware of the risks that modern drugs pose to human health, such as leaving drug residues in food, or creating 'superbugs' that are resistant to antibiotics. In organizations like Anthra, socially engaged veterinarians merge local knowledge with scientific information, playing a unique role that deserves to be emulated. The training videos made with these veterinarians and their farmer allies will hopefully show more people that it is important to bring the best of both worlds together.*

Related video

Even remote pastoralists are now making gratuitous use of animal antibiotics. In this video from Nigeria, Dr Akintayo, mentioned in this section, explains why it is important to avoid abusing antibiotics.

Agro-Insight (2016) *Keeping Milk Free from Antibiotics*. Hosted by Access Agriculture. 9 minutes.
www.accessagriculture.org/keeping-milk-free-antibiotics

See other videos in the Access Agriculture animal health category:
www.accessagriculture.org/category/133/animal-health

A version of the section above was previously published on www.agroinsight.com in 2018, by Paul Van Mele.

Caring for Animals, with Plants

The *Yapuchiris* are expert farmers on the Bolivian Altiplano who always have some new idea to share (Quispe *et al.*, 2008, 2018). Take Constantino Franco, for example, who is a *jilakata*, the highest traditional authority in self-governing rural communities.

In 2015, don Constantino began to teach other farmers about a method to treat the wounds of animals. He would gather several kinds of plants, boil them in fat and let the infusion cool. It made a salve that he could apply to the wounds of livestock.

At Promotion of Sustainability and of Shared Knowledge (PROSUCO – *Promoción de la Sustentabilidad y Conocimientos Compartidos*), the Bolivian NGO that supports the *Yapuchiris*, agronomist Sonia Laura encouraged don

Constantino to teach others about the remedy. She also wondered if it was really effective, so she asked a livestock expert, Elva Vargas, to investigate. Elva contacted a veterinarian, Sefarín Mena, who knew about the active ingredients of the plants used in salves and who confirmed the value of don Constantino's ointment.

Validating local knowledge in this way ensures that local treatments can be shared confidently with a wider audience.

In 2019, I watched as don Constantino explained his method to *Yapuchiris* and to other farmers attending a workshop held in the remote village of Chigani Alto on a hillside overlooking Lake Titicaca. *Yapuchiris* from distant communities had come to work with local farmers. They broke into groups and spent the morning on different farming topics, such as seed, weather and soil.

Don Constantino had gathered an enthusiastic group around him. His new friends from Chigani Alto went to the nearby hills and returned with a selection of medicinal plants. They ground the plants in a metal hand-cranked grinder.

The group boiled the plants in fat in a new earthen pot, to avoid adding a bitter taste to someone's good cooking pot. They strained the mixture and ladled it into little plastic containers, so everyone at the workshop could take some of the salve away with them. The experiments would continue at home.

It was a simple but valuable exercise, sharing an effective local practice that is widely available to farmers and reduces their dependency on synthetic products. Being able to make supplies instead of buying them from agro-input dealers reduces dependence on chemicals, and is important for smallholders who are often making a living on tight profit margins.*

📹 **Related video**

There is also a long tradition of local medicines for livestock in India. This video is available in English, Spanish and over 20 other languages.

Priya, S. (2019) *Deworming Goats and Sheep with Herbal Medicines*. Hosted by Access Agriculture. 11 minutes.
www.accessagriculture.org/deworming-goats-and-sheep-herbal-medicines

---

A version of the section above was previously published on www.agroinsight.com in 2019, by Jeff Bentley.

Guinea Pigs

Development planning often fails miserably. Figuring things out along the way often works better (Easterly, 2006). In this case from Peru, some creative farmers and competent outsiders were able to pull together results from different projects to cash in on guinea pigs.

Fig. 9.3. Cutting fodder for guinea pigs.

After 2013, in Quilcas, in Junín, Peru, Lucía Ávila and her neighbours started by raising a mix of fodders, including rye grass, lucerne (alfalfa) and clover, with some help from the agronomists at Yanapai, an NGO. On the down side, these grasses and legumes needed irrigation. With support from the government of Peru, the farmers of Quilcas dug an irrigation canal from some 7 km away, and the people began growing small patches of fodder which they could cut for several years, fertilizing it with ash and manure. Then the fodder patch would be dug up and planted in potatoes, which prospered in the soil where the fodders had been grown.

Every day, doña Lucía has been able to cut two large blankets full of fodder, enough for a milk cow, or in her case, enough for 200 guinea pigs (Figs 9.3 and 9.4).

Doña Lucía had started cautiously. In 2014, she got her first pair of guinea pigs from an NGO, Centre for Rights and Development (CEDAL – *Centro de Derechos y Desarrollo*). The rodents reproduce quickly, so she soon had dozens of guinea pigs. Every year she gets some new males, to avoid inbreeding. She specializes in a large, meaty breed called *Mi Perú* (My Peru), which is white and red, like the Peruvian flag.

Fig. 9.4. Guinea pigs are a popular small livestock in the Andes that provide regular income to smallholder farmers, women in particular.

As doña Lucía explains, guinea pigs give her a steady income. Plenty of customers come to her house, and she sells the guinea pigs for 20 soles (over US$5).

She says that before she got the big red-and-white guinea pigs, she had some others which she describes as 'small, like rats, and the colour of rats.' She adds, 'When you have grass you can have nice, fat guinea pigs, and you can sell them and have a little money. You can improve your standard of living.'

While guinea pigs are thought of as pets in many northern countries, in places like Peru they are small livestock. They are easy to raise at home, in the courtyard, under the shade of a porch.

Formal development is often criticized as being prone to failure. So, it's only fair to recognize its successes. In this case, three different projects happened to come together from different institutions. The canal waters the fodder, which feeds the guinea pigs (and cows). It works as a system, while (with a bit of serendipity) the parts came from three different organizations. The farmers used these innovations creatively to make a new livestock system, even if it was not part of a grand master plan.*

📹 **Related video**

This video is from Quilcas, Peru, on livestock and improved fodder.

Agro-Insight and Yanapai (2022) *Improved Pasture for Fertile Soil*. Hosted by Access Agriculture. 16 minutes.
www.accessagriculture.org/improved-pasture-fertile-soil

---

*A version of the section above was previously published on www.agroinsight.com in 2022, by Jeff Bentley.

Eating an Old Friend

In 2018 in Bangladesh, in the village of Begati Chikerbath, I visited Shamsur Naheris, an energetic extensionist in a bright orange sari. She had organized an exchange visit so that local women could tell their stories about making money and changing their lives by simply raising chickens.

A year and a half earlier, the village had hosted a farmer field school (FFS) on poultry, where the women learned to vaccinate their chickens and ducks with eye drops and to keep the hens in small coops. When the hen has a clutch of eggs, she sits on them in a nest, called a *hazol*, which the villagers make themselves, a technique they learned in the FFS. The *hazol* is a kind of earthen bowl (made with a bit of Portland cement) with two small cups on one side for feed and water. The *hazol* is big and heavy, so the hens are less likely to upset and spill their food. The hen sits on straw in the *hazol* and broods her eggs with water and food handy. The *hazol* and the hen are placed inside a small chicken coop.

More chicks live to maturity with this system, and when they are 6 weeks old, they can be let loose to find their own food, which lowers costs and saves space in the chicken coop. Then the hen can start another brood. This way she gets five or six broods in a year, over a useful life of some 5 years, until she ends up in the family cooking pot.

'How can you stand to eat your old friend?' one visitor asked, concerned that the women might have become too attached to the hens to eat them.

'It's easy, we just soften the meat first with green papaya', one of the chicken farmers explained.

While there may be little sentimentality attached to the birds, the women are all keen to raise them. Every house has a small chicken coop in the back yard, and all of the little structures are filled with healthy birds.

In a meeting with visitors from other villages, five local women told how raising chickens improved their income, and that having the extra money also boosted their self-esteem. Members of the audience soon began to ask for chicken projects in their communities as well. The visitors were farmers and their husbands, 25 couples from six local community-based water management groups. Having the husbands attend was a touch of inspiration. It would ensure that the men would be convinced and would support their wives as they started small-scale commercial poultry farming.

Even a simple technical innovation, such as a chicken coop and an improved nest, may require some training and community organizing. Farmers who take an FFS are supposed to share results with neighbours, although they do not always do so (Tripp *et al.*, 2005). Teaching others may take a little outside facilitation, such as inviting FFS graduates to appear in a video. Many of the videos referenced in this book were filmed with FFS graduates, who were delighted to address a larger audience.*

Related video

People who raise chickens at home face similar problems, and can use the same solutions, from Bangladesh to Kenya.

NASFAM, NOGAMU, Egerton University, ATC/UNIDO and Songhaï Centre (2016) *Taking Care of Local Chickens*. Hosted by Access Agriculture. 10 minutes.
www.accessagriculture.org/taking-care-local-chickens

---

*A version of the section above was previously published on www.agroinsight.com in 2019, by Jeff Bentley.

It Takes a Family to Raise a Cow

Emma had it all, including an award-winning advertising career in Nairobi, when she decided that she was tired of the rat race and wanted to spend more time out-of-doors. So, she took some training in business management and in dairy, and in 2002 she began driving out to the village some 50 km from Nairobi where her family had 10 acres (4 ha) of land, some cows and a barn.

Emma loves the cows and the dairying, but the low milk prices frustrate her, so she has a shop in a nearby town where she sells her own milk and her hand-crafted, bio-active yoghurt. She also sells dairy calves to other farmers.

Seven full-time workers on the farm tend the cows and raise the fodder: napier grass, some lucerne (alfalfa) and some maize under drip irrigation. The only problem is that thirsty antelopes come to eat the hoses, to get the water out.

Emma warmed up milk for us to drink, rather than tea. She hasn't lost the advertiser's touch; she's always promoting milk. She gave us the smoothest explanations of how to make yoghurt, even while being interrupted to talk to suppliers and customers on the phone. Emma is articulate and friendly, an easy woman to like and to respect, so I wish I could end the story here, about how a clever, educated woman loved farming enough to make it a business.

But there's another side to it. Emma lives in the city and commutes to the farm. If she's sick or has something come up, she might not make it to the farm for a few days.

Her workers avoided our gaze. It wasn't clear what they were doing, but they weren't working very hard. There was clean, running water on the farm, but the workers had not bothered to wash the manure out of the barn or fill the watering troughs for the thirsty cows. The fodder had been allowed to get wet and was spoiling in the feeding troughs. The cows were caked in dung. One cow lay in the muck, panting with a fever.

Compare this to another farm we had seen, run by Peris, a mature woman in the same part of the highlands. Peris also relied on hired workers for much of the physical labour, but her employees were busy sweeping the barn, and smiled when we caught their eye. The young men put the dung onto the compost pit as they washed the floor. Each cow had a comfortable rubber pad to stand on. There were no flies and little smell. Peris had a model cow barn.

The difference is that one of these dairy farmers lives right on her farm. Peris's back door opens onto the cow barn. When she's working with her cows she can see her kitchen window, and she sees her kids when they walk home from school. By always being on the farm, Peris can keep an eye on her herd and quickly set any mishaps right.

One of these remarkable ladies is farming like a business, but the other is farming like a family, and that makes a difference.*

Related video

Milking by hand can be a pleasant experience for the milker, and the cow, as explained in this video from Kenya.

Egerton University (2016) *Hand Milking of Dairy Cows*. Hosted by Access Agriculture. 9 minutes.
www.accessagriculture.org/hand-milking-dairy-cows

A version of the section above was previously published on www.agroinsight.com in 2014, by Jeff Bentley.

Smelling Is Believing

In Uganda, Emmanuel Ssemwanga was telling us about a self-cleaning pig pen. The manure just piles up and has no odour, thanks to indigenous micro-organisms the farmer applies to it.

Emmanuel was trying to convince Paul and me to help him make a video on what he called the 'organic pigsty'. The basic idea is to dig a pit and fill it with sawdust and leave the pig in the pit without ever cleaning it. But we weren't buying it. 'That pit will turn to a cesspool', we said. And when

Emmanuel said the idea came out of Makerere University in 2011, we said, 'This technology is too young. Wait until farmers adapt it.'

But they had. The day after we doubted Emmanuel's idea, we met local farmer Caroline Nasamba who told us about a family that was using the new pig pen, 'and the pigs are so clean they look like they just stepped out of the shower', Caroline said.

The next day, Paul, Emmanuel and I joined Caroline and colleagues James and Noel, along with John Kateregga, a friend of Caroline's, who led us to the farm.

We got off the bus in the town of Entebbe, near the capital of Kampala and also the site of the international airport. It is an example of what is now called peri-urban: half-town, half-countryside. The small houses are close together. There is little land to spare and people are trying to grow crops and gardens and raise animals, but at least they have easy access to markets.

John introduced us to Noola Nalongo and her son Waswa, who stays with his mother and tends the pigs while her six other sons and daughters attend Makerere University. There is no shortage of ambition here.

Waswa showed us the pig pen, and there was no stench at all. The pigs looked fat and happy. Waswa turned over a lump of the rich, black soil in one of the six little pens. A fat earthworm slithered back into the clod of earth. This soil was alive, and made by people and pigs.

Paul took some of the compost into his hand. 'Smell this', he told me. I held it up to my nose and breathed in the rich musty scent. 'It smells like forest soil', Paul said.

Still sceptical, we looked for flaws. The floor of one pen had a puddle in it. Waswa turned the muck over with a hoe and the water seeped into the overturned compost.

The pig house was built with scrap lumber and recycled sheet metal. It was a poor family's investment. The sty was divided into six pens.

Waswa explained that he built it by digging down four or five feet and filling the pit half full with sawdust. He then sprinkled a mix of sugar-rice-water on it, which he made himself. He adds 1 kg of boiled rice and 1 kg of sugar and a spoonful of salt to 20 l of water. He leaves it to ferment for a week, and it attracts and cultures local microorganisms, like artisanal wine gathering wild yeast. Every week, Waswa spreads 5 l of the stuff on the floor of the pig pens. Once a week he turns over the top layer of the muck in the pig pens. The muck looks almost like soil. The manure and urine mix with the absorbent sawdust and, with the help of the native bacteria in his fermented brew, the whole bed of muck composts almost immediately. (For some reason it wasn't very warm. It doesn't burn the pigs.)

Every 3 months, Waswa digs out the muck. Neighbouring farmers want it as fertilizer, so they buy sawdust and bring up one bag of it, which they trade with Waswa for three bags of compost.

Nothing smells as sweet as success. Noola and Waswa had never had pigs before. A friend loaned Noola 70,000 shillings (about US$27) to buy a sow, and the friend taught the family how to feed the pig. Waswa has since paid her back. When the sow had piglets, Waswa sold them and bought another sow. Then they had 20 piglets, which they raised and sold to pay tuition at the university.

Still, a technology isn't ready until farmers start to copy it from other farmers. This was when John, our guide, said he was planning to make his own sawdust-pit-pig-house.

John and his wife and five kids and two cows live on a plot no larger than a suburban garden. John had already made a pig house of brick and cement when he met Noola and Waswa. John tore out the cement floor, hired a neighbour to dig a pit into the ground and will soon fill it with sawdust and put a pig into it. 'What if you didn't have sawdust?' Paul asks.

'I would use grass', John says. He plans to build a goat pen on top of the pigsty and soon add another pig house next to the first one. John is making the idea his own.

Pigs are semi-aquatic, river-bank animals by nature, designed to live on soft soil. They like to dig and often break their concrete floors to get their snouts into the earth below. A sawdust pad that mimics a forest floor allows a pig to be a pig.

Emmanuel and colleagues exchanged phone numbers with John and Waswa and plan to return to include this innovation in a video, to share the idea with other farmers. It's one of the stranger innovations we have seen for a while, but shows that a bit of research and farmer ingenuity can pay off.*

📹 **Related video**

The idea of odour-free pigs is spreading around the world. In this video, Filipino farmers use good microorganisms on deep beddings in pig pens, to reduce labour, gas emissions and the need for antibiotics.

Philippine Agroforestry Education and Research Network (2024) *Raising Pigs with No Smell and Less Work*. Hosted by Access Agriculture. 13 minutes. www.accessagriculture.org/raising-pigs-no-smell-and-less-work

*A version of the section above was previously published on www.agroinsight.com in 2014, by Jeff Bentley.

References

Anthra (2023) Anthra website. Available at: www.anthra.org/ (accessed 16 October 2024).

Dutch Farm Experience (2021) Lessons Learnt in Dutch Dairy Farming. Available at: www.dutchfarmexperience.com/ (accessed 16 October 2024).

Easterly, W. (2006) *White Man's Burden: Why the West's Efforts to Aid the Rest Have Done So Much Ill and so Little Good*. Penguin Press, New York.

Fagan, B.M. (ed.) (1996) *The Oxford Companion to Archaeology*. Oxford University Press, Oxford, UK.

Ghotge, N.S. and Ramdas, S.R. (2008) *Plants Used in Animal Care*. Anthra, Pune, India.

McNeil, I. (1990) *An Encyclopedia of the History of Technology*. Routledge, London.

Natural Livestock Farming Foundation (2021) Natural Livestock Farming website. Available at: www.naturallivestockfarming.com/ (accessed 16 October 2024).

Quispe, M., Baldiviezo, E. and Laura, S. (2008) *Yapuchiris: Ofertantes Locales de Servicios de Asistencia Técnica (Yapuchiris: Local Technical Assistance Service Providers)*. PROSUCO, Cooperación Suiza, La Paz, Bolivia.

Quispe, M., Laura, S. and Baldiviezo, E. (2018) *Yapuchiris: Un Legado para Afrontar los Impactos del Cambio Climático (Yapuchiris: A Legacy to Face the Impacts of Climate Change)*. PROSUCO, Cooperación Suiza, La Paz, Bolivia.

Scott, J.C. (2017) *Against the Grain: A Deep History of the Earliest States*. Yale University Press, New Haven, Connecticut.

Tripp, R., Wijeratne, M. and Piyadasa, V.H. (2005) What should we expect from farmer field schools? A Sri Lanka case study. *World Development* 33, 1705–1720.

PART 3

Social Innovations for Healthy Food Systems

Part 2 focused on technical innovations for agroecology. Now we turn to social forms of collaboration. Farmers and their allies can organize to create shorter food chains, and to revive or reinvent local food cultures. Organic certification and polices that enable innovations such as mobile slaughterhouses can benefit family farms. Family farming is constantly changing, responding to new demands from consumers, new regulations, a changing climate and changing market dynamics. Smallholder farmers themselves also want to produce for the market, to meet their needs for medicines, clothing and school supplies for their children. Technical and social innovations can be created in part through enlightened collaboration between farmers, researchers, development professionals, as well as entrepreneurs and local authorities.

10 Farmer Cooperation

Abstract

Farmer associations empower farmers, especially women. For example, women's groups in Ecuador and Bolivia are able to gain leadership skills, pool machinery and negotiate better prices as they produce high-value crops, like dairy and vegetables. A group of organic farmers in India uses digital tools for planning, marketing and home delivery of produce. A group of Kenyan women have been able to solve their disagreements about pricing and labour contributions, and successfully sell banana flour. Young organic farmers in Belgium collaborate on seed production, knowledge sharing and market diversification. Farmer groups may need outside facilitation at first, but organization helps farmers to overcome challenges, improve livelihoods and build a more sustainable food system.

A Safe Space for Women

In Ecuador, as in many other countries, the rural exodus of men to cities and abroad, in search of work, imposes new duties and responsibilities on the women who are left behind. Besides having to look after their households, they now are also left to manage the farm alone and sell their produce. For indigenous women, this poses special challenges because so many men have left, and until recently the women held few formal positions of leadership. Without the necessary confidence and proper skills to negotiate prices with middlemen, these women continue to be exploited and forced to live in poverty.

As Marcella, Jeff and I visited the NGO EkoRural in Quito in 2022 to meet with the director, Ross Borja, and scientific advisor, Pedro Oyarzun, it was encouraging to hear how things are gradually changing thanks to many years of support by local organizations, backed by a growing pressure of the international community to advance women's rights and consider gender in all their activities. For rural women to gain confidence and control over their lives, getting organized into women's associations is crucial. To nurture future women leaders, this can best happen when they are given a safe space.

When women grow food without agrochemicals this makes them stand out from other food sellers in the cities, and many urban clients are willing to pay an extra price for healthy food. 'But without the necessary skills and loaded with the historic mistreatment, indigenous women have little chance in conventional food markets to sell their produce at a fair price', says Ross.

She explains that for a decade, EkoRural has been helping indigenous and mestizo smallholder farmers to sell their produce and build a client base in urban centres by establishing new dedicated agroecological markets at family planning centres. 'As only women with children come to the health centres and most of the doctors are also women, this has offered a relaxed environment for indigenous rural women to gradually develop their skills and confidence', says Ross. Once a week, the women set up their stalls in the garage of the family planning centres.

EkoRural has also supported workshops where the women doctors together with their patients and farmers learn about food and nutrition. The link between healthy food and human health is an obvious one, but using urban family planning centres to create a market for disadvantaged rural women is something I had never heard of before: truly an innovative approach that at the same time helps create a relationship between food producers and consumers.

That this approach is bearing fruit has been proven by the past 2 years of the COVID-19 pandemic. While the lockdown closed down these new markets, the urban consumers, including the doctors, have established such good relations with the rural women that they continue to buy their produce by placing orders on the phone or social media. Not all of the indigenous women are familiar with digital communication, but their children are.

Another example of EkoRural's concern for rural women is the initiative to sell agroecological products at the Casa de la Mujer in Riobamba. This initiative not only allowed them to sell but also enabled young rural women to come into contact with gender issues and leadership at the provincial and national levels and to experience leadership roles in this organization.

When attention to gender is taken seriously and there is the necessary creativity and investment in building rural women's organizations, it is possible to establish alternative food networks. Gradually and in a supportive environment, indigenous women can gain the necessary skills and confidence to earn a decent livelihood in the absence of men.*

Related video

Indigenous women in Ecuador have strengthened their associations with the support of a local NGO and local authorities.

Agro-Insight and EkoRural (2022) *Inspiring Women Leaders*. Hosted by Access Agriculture. 14 minutes.
www.accessagriculture.org/inspiring-women-leaders

A version of the section above was previously published on www.agroinsight.com in 2023, by Paul Van Mele.

What Is a Women's Association About?

In Ecuador, community organizer Ing. Guadalupe Padilla has told me that belonging to a group can help women gain leadership experience. Women become leaders as they work in a group, not in isolation. Guadalupe has helped to organize several women's groups. A group has to have a purpose, and that purpose can easily be related to agriculture.

In Cotopaxi, Ecuador, in 2022, while working with Paul and Marcella from Agro-Insight to film a video on women's organizations, we met Juan Chillagana, vice-president of the parish (town) council. As an elected local official, Mr Chillagana mentored several women's organizations, each one organized around a specific product. We caught up with him on 4 February as he met with a group of women who were growing and exporting golden-berries (*Physalis peruviana*). The fruit buyer was there, wearing a hair net, which gave him the air of a serious food dealer. He said, 'All we ask is that you don't apply agrochemicals.' The association members and the buyer weighed big, perfect goldenberries in clean plastic trays to take to the packing plant.

We talked with one of the farmer members, Josefina Astudillo, who seemed pleased to be trying this new fruit crop. She guided us to her field, about a kilometre from the community centre where the meeting was held. Doña Josefina proudly showed us her field where the fruit was ripening to a golden perfection. One woman could grow goldenberries by herself, but it takes a group to meet the buyer's demand: 1000 kg of export-quality fruit per week.

We also met Beatriz Padilla (Guadalupe's sister), a small-scale dairy farmer, who leads a group of 20 households who pool their milk. The association sends a truck to each farm, collects the milk in big cans and transfers it to the group's cold tank. Twice a day, about 1500 l of milk are collected by two different buyers, including one who comes at 3 AM. It's a lot of work. Doña Beatriz explained that she couldn't do it without the group. She needs the other families so they can get a better price for their milk. A farmer with two cows has to take whatever price the dairy will give her. But an association can negotiate a price.

Margoth Naranjo is a woman in her 60s who has worked her whole adult life in associations, often in groups that included men as well. She started in her local parent-teacher association, helping to organize the children's breakfast. Later, she was the secretary of a farmers' insurance group, until she became the treasurer and then the president. Now, with the Corporation of Indigenous and Peasant Organizations of Cusubamba (COIC – *Corporación de Organizaciones Indígenas y Campesinas de Cusubamba*), doña Margoth is helping several women's organizations, which sell their own agroecological vegetables, to band together for added strength. This work came to a standstill during the COVID-19 lockdown, but doña Margoth started organizing again in 2022.

All of the women's leaders we met in Ecuador were part of a group. And each group was formed for a concrete purpose, whether for goldenberries, milk or vegetables. Each specific purpose was linked to a dream the women's groups share: to have a quality product to sell, to improve their livelihoods. The women improved their leadership skills with practice. They said that any woman could be a leader if she joined a group and participated long enough.*

Related video

The above experience is documented in this video.

Agro-Insight and EkoRural (2022) *Inspiring Women Leaders*. Hosted by Access Agriculture. 14 minutes.
www.accessagriculture.org/inspiring-women-leaders

A version of the section above was previously published on www.agroinsight.com in 2022, by Jeff Bentley.

Look Me in the Eyes

In Ecuador in 2022, I saw some of the best extension work I have ever seen. Fernando Jácome, an agronomist with SWISSAID, took us to meet farmers, almost all women, who have been working with him and his colleagues for over 10 years. Eighty-five smallholders from different communities of Pelileo, in Tungurahua in the Andes, are organized into seven small associations. They have learned to produce an impressive assortment of fruits and vegetables, from tomatoes to strawberries, cabbage, lettuce, avocados, lemons, blackberries and many more, as well as rabbits and guinea pigs. It's all grown ecologically.

With Paul and Marcella, from Agro-Insight, filming a video on agroecological fairs, we accompanied Ing. Alex Recalde, an agronomist working for the Pelileo municipality, as he inspected farms to make sure that they were really producing ecologically. Alex's visits were largely about teaching and encouraging, with little policing, since the women all seemed convinced about agroecology.

First, Alex registered what the farmers were growing. That way he knew what each one would harvest, to later verify in the fair that they were only selling their own produce, and in plausible amounts.

During the farm visits, often accompanied by leaders of the Agroecological Associations of Pelileo, Alex looks for signs of chemicals, such as discoloration on the leaves, or residues of synthetic fertilizer on the soil, or discarded chemical containers. He also looks at the insects on the farm. A diverse insect community with many beneficial invertebrates and few pests is a sign that toxic chemicals have not been used.

If the farmer has any pests and diseases, Alex advises her on what to do. We were with him while he explained to farmer Korina Quille that the unsightly scabs on her avocados were not actually a disease at all, but were simply scars formed because the wind had rubbed the tender fruits against a branch. Realizing that cosmetic damage is not caused by a pathogen can also reassure farmers that agroecology is working for them. It also helps them explain to customers that there is nothing wrong with their avocados.

Later that afternoon, we attended a meeting of the agroecological association. The organized women began by taking attendance (roll call). They had brought samples of their produce for an exercise on displaying it attractively and in standard-sized pots and baskets, so they could all sell the same measure at the same price, one that would be fair for farmers and consumers.

SWISSAID's Mario Porres led a lively discussion, asking the audience: 'How can you have a standard measure, if the customers all insist on the *yapa* [a little bit extra]?' He held up a basket of berries and said, 'Measure it, take a few out, and when the customer asks for the yapa, put them back in.' The audience laughed in appreciation.

The meeting ended with a drama coach, Verónica López, who used theatrical exercises to build the women's self-confidence. Poor, peasant and indigenous women can be afraid to be assertive, but Verónica was teaching them to be bold and to have fun at the same time. The women knew Verónica, and as soon as she took the floor, everyone stood up. 'Walk angry!' Verónica shouted, 'your husband has been telling you what to do!' The women stomped around the courtyard, arms swinging, recalling their anger, over-acting and loving every minute of it.

'Now, imagine that you bring that anger to the market, and you are angry with the customers. Will they want to buy from you?' Verónica asked.

In another exercise, on love, the women hugged each other, and they learned to walk happy, not angry. The drama coach also had the women shout, part of an exercise where they learned to speak loudly, but kindly, looking customers in the eye to win them over.

This training, encouragement and organization has opened a space where indigenous women can sell their beautiful produce in the local open-air markets, in small cities like Pelileo, and in big ones like Ambato.

Later, we found out how well the training had paid off. One morning before dawn, I was with Paul and Marcella in the wholesale market in the city of Ambato. This is the biggest market in Ecuador, a sprawling complex of open-air pavilions, where trucks loaded and unloaded produce. Fernando Jácome, the extensionist, had brought us here, to the heart of the country's commercial food system, but he left us for a while with Anita Quille, one of the women leaders of the association. When a local official approached us to ask why we were there with a big camera, doña Anita stepped forward and looked him in the eye. She spoke gently but firmly, in a self-confident tone of voice, explaining who we were, and that we were there making a video on local farmers and markets.

Teaching farmers to farm ecologically is a good start, but it must also include training in organization, marketing and even assertiveness, if the farmers are going to have products to sell.*

◼️🎥 Related video

This is the video we made in Pelileo, Ecuador.

Agro-Insight and SWISSAID (2022) *Creating Agroecological Markets*. Hosted by Access Agriculture. 16 minutes.
www.accessagriculture.org/creating-agroecological-markets

---

*A version of the section above was previously published on www.agroinsight.com in 2022, by Jeff Bentley.

Home Delivery of Organic Produce

The financial crisis in 2007–2008 exposed the vulnerabilities of a global market system that put blind faith in the financial institutions. COVID-19 has been another wakeup call: the more one depends on global trade, the more fragile local food supply becomes. In many countries, weekly markets closed down, and while the need for food had not diminished, many farmers struggled to harvest and sell their produce.

Online shopping got a boost across the world. A video produced by one of our partners in India in 2022 shows that when farmers are organized, they can combine the best of both worlds: build on the trust they established with clients during face-to-face interactions, and use digital marketing tools to promote and sell their farm produce.

The video shows how some organic farmers in Maharashtra, India, established a group, called Saad Agronics. They have between 80 and 200 clients a week who place orders via a digital platform. Customers receive farm produce, guaranteed to be fresh and healthy, at their doorstep, while the farmers get a higher price than they would earn in the market.

For a single farmer, it is not profitable to take a few vegetables to the homes of urban consumers. As Jalindar Jadhav, one of the group members, explains in the video: 'Previously, I started a home delivery service for my organic farm produce. But consumers want different types of vegetables and other food. It was not possible for me to grow all types of vegetables. So, with only a few products to offer to customers, this home delivery service was not profitable for me.'

Before each planting season, Afrin Kale organizes a meeting with 25 of her fellow organic farmers to plan who will grow what type of vegetables and other produce for the group. This avoids duplication and ensures that there is enough diversity to cater to the broad demands of their clients (Fig. 10.1).

Fig. 10.1. Organic home delivery services require farmers to cooperate with each other. Photograph used with permission from Access Agriculture.

'We plan in such a way that two farmers in our group never grow the same vegetables and each of us practises crop rotation. At any given time, a farmer can grow three, four or more types of crops. Not all farmers grow vegetables, some grow fruits, and some grow pulses and different rice varieties or other cereals. We also check how much water is available for each farm and for how many months each farmer can supply produce', says Siddhesh Sakore, another group member.

While COVID-19 increased the demand for healthy organic food and for home delivery services, the organized farmers at Saad Agronics were ready to respond to these changing demands. They continue to strengthen relations with their home delivery customers through open farm visits and by using social media to expand their client base. To collect orders and delivery addresses of clients, the group also had an app developed (Fig. 10.2).

As with any home delivery system, one has to avoid spending all the profits on fuel and labour costs. Saad Agronics decided to only deliver in parts of the city where they had a minimum set of orders. The young members of the group are very savvy with digital tools: they use GPS to ensure fast delivery using the shortest road, they use WhatsApp to inform clients of the time of delivery and they have an e-wallet so clients can pay with their mobile upon delivery.

Many organic farmers want to join this group, so the future looks bright for farmers and consumers who are all concerned with healthy food, free of chemicals. Hopefully, this video from India will inspire farmers across the world.*

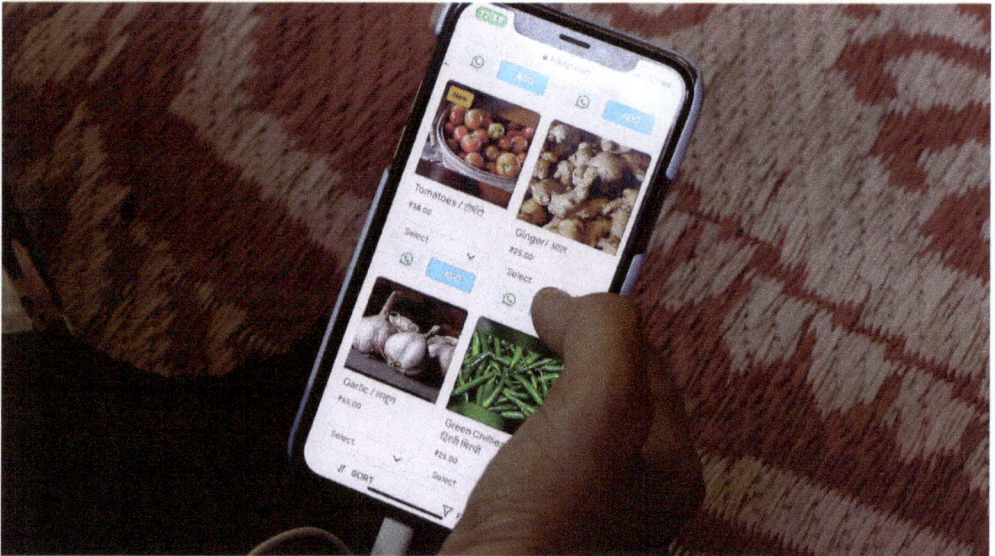

Fig. 10.2. Some farmer groups invested in a mobile ordering system. Photograph used with permission from Access Agriculture.

Related video

During the COVID-19 lockdown, farmers in India organized to deliver organic produce, which consumers ordered online.

Pagar, A. (2022) *Home Delivery of Organic Produce.* Hosted by Access Agriculture. 15 minutes.
www.accessagriculture.org/home-delivery-organic-produce

A version of the section above was previously published on www.agroinsight.com in 2022, by Paul Van Mele.

Gardening Against All Odds

All over the tropics, from Lima to Lagos, the big cities are bursting with migrants. In some regions, like the Andes, I have seen parts of the countryside emptying out, with whole villages boarded up.

Rural-urban migration is often seen as a problem, fostering crime and underemployment, straining city services (Mlambo, 2018) and leading to mental health problems (Zhong *et al.*, 2018). Moving also costs migrants money and creates distance from the rural homes that they love, but often allows people to make more money than they would have made in the countryside (Selod and Shilpi, 2021). See also the section 'Choosing to Farm' (Chapter 11, this volume).

In 2019, my wife, Ana, and I visited a neighbourhood of migrants on the edge of Cochabamba, a city that has long been divided into a fashionable north side, hemmed in by mountains, and a working-class south side. But in the past 15 years or so the south side has mushroomed out of the valley bottom, to grow over the hills south of town. At night, the lights on the hills are a reminder of how much the city has changed.

In one of the newest of these poor neighbourhoods, we met some of the 80 members of a women's group, *Nueva Semilla* (New Seed). Migration has been intense after the mining industry crumbled in the 1980s, but even since then, people have continued to leave villages in the provinces of Cochabamba and in northern Potosí, the poorest region of Bolivia, to seek a better life in the city (Goldstein, 2004; Delgadillo Camacho *et al.*, 2018).

Nueva Semilla is in a tough neighbourhood where people have to look after themselves. Families live on small plots of land, where they slowly build their brick and cement houses with their own hands in their limited free time, usually just Sundays and national holidays. The streets are unpaved and dusty, laid out on square grids (or in curves on some of the steeper slopes). The government has built schools and hospitals. There is electricity, but no running water. People buy water from tanker trucks for a dollar a barrel.

Nueva Semilla started in 2014, when some of the women were taking a catechism class. They were impressed with the garden in the churchyard and this set them thinking. They had all been farmers in the places they had come from; why not establish their own gardens in their new homes?

But the women were used to growing potatoes, maize and barley, not garden vegetables. Fortunately, an NGO, *Agroecología y Fe* (Agroecology and Faith), helped them with seed and some training, and some fabric to make semi-shade to protect the young plants from the fierce sun.

Doña Betty, one of the leaders, showed us her house plot, a small square of rocky hillside with no soil. Doña Betty bought a truckload of loamy soil, which she mixed with leaf-litter she collected from beneath mesquite trees on the surrounding hills. She put the mixture in old tires and irrigated it with water she bought from the cistern truck. She has created a delightful garden, with a dozen different vegetables, including healthy, organic tomatoes and celery which she is growing for seed to share with the members of her group.

We paid a small fee, along with a small group of other visitors, for lunch which the women made. They were eager to sell their vegetables. Four heads of lettuce went for about US$0.65, cheaper than in the market. The families eat a lot of their own produce and the kids we saw appeared healthy and well fed. The women's small vegetable gardens were surprisingly productive, even if they have to make their own soil and buy their water. The families even have surplus produce to sell.

The NGO is planning a seed exchange fair too. Once a month they also have a solidarity fair, where the women sell 'solidarity baskets' of vegetables they produce themselves.

The women and their families have left their farms behind, but they have also brought the best of country values with them: hardwork and creativity. These adaptive people have taken their personal development into their own hands and have decided that a job in the city and a home garden are tickets out of poverty.*

◼ Related video

Farmers in Nepal have learned how to manage tomato late blight ecologically.

Practical Action Nepal (2019) *Managing Tomato Late Blight*. Hosted by Access Agriculture. 9 minutes.
www.accessagriculture.org/managing-tomato-late-blight

A version of the section above was previously published on www.agroinsight.com in 2019, by Jeff Bentley.

The Red Bucket

In 2019 I had a chance to visit some dairy farmers near Cochabamba. They live in a small community and are members of a dairy cooperative which was able to buy a refrigerated milk storage tank with support from the Bolivian government. Twice a day the farmers bring their metal milk cans to the collection centre, a small brick building which houses a 1000-l storage tank.

The stainless steel tank has an electric cooler to chill the milk and a paddle that gently stirs it. This keeps the milk fresh until a tank truck from the dairy collects the milk later in the day. After milking the cows at home, the farmer simply takes her milk to the centre, avoiding the work of selling it door-to-door or of making it into cheese.

The farmers are organized in groups of a dozen or so households, and they take turns running the collection centre. This involves measuring the density of each delivery of milk with a little gadget that looks like a pistol (a density meter) to make sure that no water has been added, and writing down how many litres each person brings in.

Every 2 weeks the cooperative pays each farmer for the milk they delivered. It sounds simple, but calculating the volume of milk each farmer delivers is a little complicated.

The farmers bring in one or two milk cans each time they come. The factory that makes the milk can labels each one '40 litres' but they only physically hold 39 l. The staff at the cooperative are not sure why this is. The farmers at the collection centre have been known to naïvely give a neighbour credit for 40 l, because the can looked full. Besides, the cans are not always full, so the milk from each family has to be measured accurately, in a special pail. Pouring the milk into the pail (while trying not to spill any) is a tedious task, and another transaction cost. But it has to be done accurately. The dairy and the cooperative will fine the farmers if they report more milk than they deliver.

Another problem is that farmers report whole litres to the dairy, often rounding down actual volumes.

At the meeting I attended, one young farmer complained bitterly about this. 'Sometimes I bring in almost 5 l, and they write down 4!'

She went on to say that sometimes the person in charge is nice, and gives her credit for 5 l, but most of her fellow farmers won't do that. She singled out one other farmer, doña Irma, as being especially strict.

But doña Irma had a solution for that. 'That's why we have the red bucket', she politely reminded the group. 'If someone has a little extra milk, they pour it into the red bucket. If someone needs milk to make up a litre, they can take it from the red bucket.'

Transaction costs can be higher for smaller producers. It may take as much time and effort to deliver 40 l as to bring in 400. It is easier to deliver milk to a collection centre than to make cheese. But the collection centre adds a few new costs, such as the time it takes to run the centre, and the risks of mis-measuring the milk.

The young farmer was still angry. No doubt it is more fun to take milk from the red bucket than to add milk to it. Still, the red bucket was a local, if imperfect, solution to a nagging transaction cost.

Smallholders will make marketing and institutional innovations, like the red bucket, to stay profitable in a world where food systems are getting ever more complex. At a time when many people are leaving the countryside, and multinational corporations are monopolizing the food supply, it's good to know that at least some cooperatives are trying to work with smallholders so they can earn a decent living in their home communities.*

◼️ Related video

Milk collection centres face similar challenges across developing countries, and can use similar solutions, as seen in this video from Nigeria.

Agro-Insight (2016) *Taking Milk to the Collection Centre*. Hosted by Access Agriculture. 13 minutes.
www.accessagriculture.org/taking-milk-collection-centre

A version of the section above was previously published on www.agroinsight.com in 2019, by Jeff Bentley.

The Next Generation of Farmers

Whether in Europe or in the Global South, young farmers, unless they are born into a farm family, often lack three key things: land, money and knowledge. But a new breed of farmers has risen, fuelled by passion to produce food in a healthy way, free from agrochemicals. Their journeys are often difficult, but with support from the community and by helping each other, they are heading towards a fairer and brighter future, as I learned in 2021 on a revealing road trip.

I joined my friends Johan Hons and Vera Kuijpers on their weekly trip to deliver and buy organic produce from wholesalers and fellow farmers to stock up their farm shop that opens from Friday afternoon until Saturday noon. Johan and Vera are pioneer organic farmers in north-eastern Belgium.

'When we started some 30 years ago, it was just us and one other family who had a basic food packaging machine. Whenever we needed, we could use their machine', Johan said. In the meantime, the number of organic farmers has grown, and an amazing informal network is coming to life.

The back of the van is loaded with freshly harvested potatoes, a few crates of cabbages and leek seedlings that Johan and Vera reared for the new season. Having left their farm before 6 AM, by 8 AM we finished our first delivery. Biofresh, a main organic retailer, bought their potatoes. At the same time, we collect the produce they had ordered online a few days earlier. Vera guided me through the warehouse, explaining how the whole system works.

I see crates of organic pineapples from Côte d'Ivoire, bright mangoes from Ghana, ginger from Peru, fava beans, artichokes and oranges from Italy, and various local products, including Johan and Vera's potatoes.

'At first, our name was mentioned on the label', Vera says, 'but they have now replaced our name with a number, so people no longer know who has produced them. I think it is to protect themselves from their competitors.' This may well be the case, but as we continue our road trip it dawns on me that the effectiveness of this strategy may only be short-lived.

As we load the van in the parking lot, we are approached by Floriaan D'Hulster, a young organic farmer we had just met in the warehouse. He has come to buy Johan and Vera's crates of cabbages. Floriaan hands over a little cardboard box. My curiosity triggered, Johan proudly opens it and shows me the little seed packages.

> This is from our group of farmers with whom we started to produce veg-
> etable seed. The seed has been cleaned, nicely labelled and packaged at
> the premises of Akelei, the organic farm where Floriaan works, and will be
> available in our farm shop as of tomorrow.

Johan smiles. Their non-profit association, *Vitale Rassen* (Vital Varieties), was formalized in 2019 and includes organic farmers across Flanders who produce seed to EU organic standards.

On to the next destination. Like Biofresh, Sinature is a wholesaler, but they also have their own greenhouses behind their warehouse. 'We like to buy as much locally produced food as possible', Vera says, 'as that is in line with our philosophy. Many clients also ask me about this.'

As we walk through the warehouse, Vera carefully goes over various lists. I learn that Johan and Vera are also buying produce for other fellow farmers. 'Many of us have started to sell our produce ourselves directly to consumers, whether at farm markets or farm shops', Vera says, 'and it is good to be able to offer clients a rich diversity of food on top of your own produce. As we have a van and a trailer, we provide this service to our fellow farmers against a small fee to cover our costs.'

Many of the new generation of farmers have managed one way or the other to secure some land. To gain knowledge and become professional growers, for the past 20 years, the non-profit organization Landwijzer has been offering both short and 2-year courses on organic and biodynamic farming.

The remainder of the day we make various stops to buy and deliver fresh produce at some inspiring farms. As Johan and Vera are pioneers, they know everyone involved in the organic food system. Many of the new generation of farmers have also done their internship with them as part of their Landwijzer course, so they have a strong bond. By providing this weekly service, they also get a chance to chat with their colleagues and exchange ideas and recent news.

When I ask Johan how the new generation of farmers is coping with the purchasing power of large buyers that push down prices, he explains that price formation and market diversification are key aspects covered in the courses offered by Landwijzer.

A few days earlier I had an online meeting with one of the coordinators of the Fairtrade Producers' Organization from Latin America. To secure a living income, cocoa and coffee growers are also forced increasingly to look at income and market diversification. While the food industry may gradually come to realize that paying a fair price for food is needed to keep farmers in business, it is reassuring to see that farmers continue to innovate by proactively strengthening ties between themselves and the community of consumers. Belonging to a network may make a vital difference for new farmers, who often lack land and a family connection to agriculture.*

📹 **Related video**

As mentioned in this section, organic farmers can organize to conserve local crop varieties.

Agro-Insight (2023) *Collecting Traditional Varieties*. Hosted by Access Agriculture. 15 minutes.
www.accessagriculture.org/collecting-traditional-varieties

A version of the section above was previously published on www.agroinsight.com in 2021, by Paul Van Mele.

Weight Watchers

A lot of donor aid is spent on strengthening farmer groups, from the local to the regional level. Another fashionable topic is value chains. Within farmer groups, tensions inevitably emerge sooner or later.

The Maisha Bora women's group in Tinganga village, central Kenya, is no exception. They started in 2012 with 15 members, but 2 years later only ten members remained.

'We have developed clear rules to which all members have to abide', says the chairperson, Margaret Ruhiu. 'From the initial group some members left. We are stronger now, as the remaining members all collaborate well.'

When probed about the type of tensions they have faced over the years, the women explain how initially they bought bunches of bananas from group members and from nearby farmers. Some women brought small bunches, others large ones, and all would get paid the same price, no matter the size.

I have seen many examples of farmer cooperatives and other groups that failed as soon as the outside facilitator stopped visiting the group. In part, this is because the outsider can be an honest broker who helps to mediate conflicts as they come up. Once that support is gone, the group dissolves into bickering. In this case, the women were able to come up with their own ways of easing tensions.

For the Maisha Bora women's group, the size of the bunches varied so greatly that it created tensions. But this is now history: they now buy bunches by the kilogram after they bought a scale to weigh the bananas.

When individual members sell bananas to the group, they may be paid for the bananas the week after, once the banana flour has been sold. Other farmers are paid on the spot.

Contributions in labour are another potential area for tension. But the group also found a solution. For processing the bananas into flour, all members contribute labour equally. The time that they work is properly recorded, so that there is no space for misuse.

The money the members make by selling banana flour is put into a bank account and supplements their weekly group savings. Each year, every member receives an equal dividend.

Trust can go a long way, but simple tools (such as weighing scales), proper rules and recording add just enough objectivity to discourage free-riders and to reward everyone fairly for their contributions. Organized groups of smallholders can function smoothly, but the members have to find ways to keep everyone honest.*

Related video

The Maisha Bora women's group described above have made a successful business, processing and marketing banana flour, as documented in this video.

KENAFF, Farm Radio Trust Malawi, UNIDO Egypt and Farmers Media (2016) *Making Banana Flour*. Hosted by Access Agriculture. 12 minutes.
www.accessagriculture.org/making-banana-flour

A version of the section above was previously published on www.agroinsight.com in 2015, by Paul Van Mele.

References

Delgadillo Camacho, M.F., Farjat Bascon, S. and Pardo Saravia, R. (2018) *Migración Interna en Bolivia (Internal Migration in Bolivia)*. UDAPE (Unidad de Análisis de Políticas Sociales y Económicas) and INE (Instituto Nacional de Estadística), La Paz, Bolivia.

Goldstein, D.M. (2004) *The Spectacular City: Violence and Performance in Urban Bolivia*. Duke University Press, Durham, North Carolina.

Mlambo, V. (2018) An overview of rural-urban migration in South Africa: Its causes and implications. *Archives of Business Research* 6, 63–70.

Selod, H. and Shilpi, F. (2021) *Rural-urban migration in developing countries: Lessons from the literature*. Policy Research Working Paper 9662. World Bank, Washington, DC.

Zhong, B.-L., Liu, T.-B., Chan, S.S.M., Jin, D., Hu, C.-Y. *et al.* (2018) Common mental health problems in rural-to-urban migrant workers in Shenzhen, China: Prevalence and risk factors. *Epidemiology and Psychiatric Sciences* 27, 256–265.

11 Shorter Food Chains

Abstract

Supermarkets dominate the food market, offering farmers lower prices, driving out smaller shops and sacrificing local sourcing and quality. One of the authors (Paul Van Mele) shares his father's experience, losing his own grocery store due to supermarket competition. Consumers and local governments have the power to support family businesses, farm shops and farmer markets from France to Ecuador. In Bolivia, Alicia García shows that a young, university-educated woman may choose to stay in agriculture, producing agroecological vegetables, while Olivier Clisson and Lisa O'Beirne are biodynamic 'baker farmers' in France who share their knowledge, passion and products with a steadily growing food-conscious consumer base. Time-honoured social constructs, such as allotment gardens, and more recent innovations, such as the Belgian 'Juice Mobile', also provide fresh food, social interaction, exercise and a connection to nature.

A Market to Nurture Local Food Culture

In the section *'La Tablée'* (see Chapter 12, this volume) I write about what can happen when local authorities support local food producers. *La Tablée* was a one-off event to inform and build a relationship with urban consumers, but there can also be more formal ways of supporting local producers.

Every city in Europe, and in many other places, has its fresh market, with food being brought in by traders and occasionally some farmers. The food being sold can come from anywhere. Not so in Rennes. This historic capital of Brittany in France has taken a bold decision to promote local food as much as possible.

On Saturday mornings people flood to the *Marché des Lices*, happily strolling in on the many narrow cobblestone streets, like bees drawn to a field of flowers. My wife, Marcella, and I immediately realize that this is not like any other fresh market.

© Jeffery W. Bentley and Paul Van Mele 2025. *Agroecology in Practice: From Local Initiatives to Global Scaling Through Video* (J.W. Bentley and P. Van Mele) DOI: 10.1079/9781800628793.0011

The diversity of produce is striking. We see market stalls with plenty of leafy vegetables and fresh herbs; others sell homemade pear and apple juice, along with cider, a traditional fermented apple beverage that is currently enjoying an amazing revival. For decades, cider had lost popularity because it could not compete against the strong marketing campaigns of the wine and beer industry. Now, this tart, fruity drink is served in every bar and restaurant. Here, in the Rennes market, a bubbly cider that will pop the cork out of the bottle is presented with a certain prestige, as if it were champagne.

There is a great atmosphere, and many people have a quick chat with the person selling the produce, whether it is cheese, a diversity of fish and other seafood, freshly chopped meat, mushrooms and other fungi, walnuts, fruit or vegetables.

One stand catches my eye. I take a closer look at one of the breads on display and ask the man what it is made of. 'Buckwheat', he says. 'This is grown locally and contains no gluten, in case you are allergic. You can keep the bread for days without it going stale.' We buy a chunk of this dark bread, which he wraps in paper. Despite the word 'wheat' in its name, buckwheat (*Fagopyrum esculentum*) is not a cereal but a member of the knotweed family (Polygonaceae) grown for its grain-like seeds.

Buckwheat grows on light, well-drained soils and has also been a traditional crop where we live in Limburg, in north-eastern Belgium, but I had never seen bread made from it. On special occasions and in a few restaurants one can eat buckwheat pancakes in Belgium, but in this part of France it is on every single menu, presented as a local specialty. During the week that we spent in Rennes, we discover various restaurants that sell only buckwheat pancakes, often served with a jar of cider.

Before we leave the market, we pass through a narrow street lined on both sides with food trucks. All of them serve buckwheat pancakes. There is the option to add a sausage, with mustard or another sauce, and then it is rolled up and eaten like a hot dog. It strikes us that the least flashy food truck has the longest cue. It is the only one that serves *galettes* made from organic buckwheat!

Later on, I learn that during the COVID-19 crisis, local producers were being pushed out of the market by middlemen and that the Rennes city council committed to changes after Olivier Marie, a respected Breton culinary journalist, made a case to boost the presence of local farmers and artisans and curb unethical merchants!

Because farmers can sell so much of their produce at this weekly market, it seems to give them enough added income to stay in farming.

Local authorities can stimulate local food production and local food processing by creating a favourable market environment. Overly rigid food regulations, as we have in Belgium, leave no space for initiatives that allow local food culture to flourish. It is time we take a better look across our borders and learn from inspiring examples across the globe.*

📹 **Related video**

Fresh, white 'farmer' cheese is part of many local food cultures.

Nawaya (2017) *Making Fresh Cheese*. Hosted by Access Agriculture. 13 minutes. www.accessagriculture.org/making-fresh-cheese

A version of the section above was previously published on www.agroinsight.com in 2021, by Paul Van Mele.

Choosing to Farm

Growing up on a mixed dairy farm in Sacaba, Bolivia, Alicia García was always interested in agriculture. In 2021, Alicia and her sister grew winter tomatoes in their new greenhouse. But as the temperature dropped near freezing several times, the plants 'burned' or died back. Alicia admits that the first winter was a learning experience.

In Cochabamba tomatoes are a summer crop, so Alicia was surprised by the cold damage, but she is sure that next winter she will manage better. To keep learning, she left one row of the damaged tomatoes standing, to see if they could recover, but she has replanted most of the greenhouse with lettuce and other leafy greens. Aphids are tomato pests, but Alicia manages them with homemade sulfur-lime and an ash-and-soap blend. Alicia fertilizes the soil with manure from her family's dairy herd and with biol (made from manure fermented in water).

As another innovation, Alicia is growing apples as an agroforestry system. Alicia planted her apple seedlings a year and a half ago, and while they are still small, she grows broad beans, onions, broccoli and cabbage in between the little trees. This makes use of the land, and keeps down the weeds.

She's also had some help along the way. When she was just 13, she began taking farming classes from the Centre for Technical Teaching for Women (CETM – *Centro de Enseñanza Técnica para la Mujer*). For the past 10 years, AGRECOL Andes (an NGO that promotes agroecology) has helped Alicia and other farmers to sell their ecological produce in coordination with the municipal government. In 2020, Alicia and her sister built two greenhouses, with support from a government programme, the Rural Alliances Project (PAR – *Proyecto de Alianzas Rurales*).

This experience shows that a young woman can be interested in agriculture enough to assume long-term commitments like a greenhouse and an apple orchard. Alicia has a lot in her favour: institutional support for training, investment and marketing, a family that provides land and manure, and she lives in an attractive community. The family home is just past the edge of the small city of Sacaba, which has all the basic services (like banks, hospitals, shops and markets). And Sacaba itself is a half-hour drive from the big city of Cochabamba. In Bolivia, rural migration is draining the countryside, but small cities like Sacaba are growing rapidly. The city also offers opportunities for farmers. Every Friday, Alicia and other farmers meet at a city park in Sacaba to sell produce to local people.

I asked Alicia why she had gone into farming. I thought she might say to make money. She surprised me a bit when she said, 'What I like is the chance to work with nature.' In other words, a lifestyle decision. She finds the work enjoyable, and she likes to farm without chemicals. Alicia explained, 'My parents never used pesticides on their farm. Even when the neighbours sprayed their maize and potatoes, my parents didn't.'

Alicia is now in university and has a year left to finish her degree in architecture. After graduation she would like to open her own office and go into landscaping, combining architecture with her love of plants and the outdoors.

Alicia doesn't farm like her parents did. They didn't grow vegetables or fruit trees, but she builds on their experience and, with appropriate help, was able to start a greenhouse and an orchard while still attending university. Agriculture can capture the imagination of the best and brightest young people.*

◼️ Related video

There are many ways smallholder farmers can intercrop short-term crops like cereals or vegetables with fruit trees, to make more efficient use of land.

DEDRAS (2018) *Growing Annual Crops in Cashew Orchards*. Hosted by Access Agriculture. 9 minutes.
www.accessagriculture.org/growing-annual-crops-cashew-orchards

---

A version of the section above was previously published on www.agroinsight.com in 2021, by Jeff Bentley.

When Local Authorities Support Agroecology

While making a training video on agroecology fairs with SWISSAID, Jeff, Marcella and I have the chance to meet the young, well-spoken Leonardo Maroto, mayor of Pelileo, a city in the Ecuadoran Andes with a population of over 50,000. Together with three women who manage the association of agroecological producers, we were given 30 minutes, which ends up becoming nearly an hour of inspiring interactions.

The initial agreement between the mayor's office and the farmers' association was signed by his predecessor, but when Leonardo Maroto explains his vision about why and how they support this initiative, it dawns on me that support for smallholder and indigenous farmers can be quite diverse.

For example, once a week the agroecological farmers hold a fair. On Wednesday, the day before the fair, we witness the city's fire brigade hosing down the covered market square. Some 16 women from the group have joined, as a *minga*, or communal work which is part of the Andean tradition. This *minga* is part of the agreement between the mayor's office and the women's association to jointly clean the marketplace twice a month.

One other commitment of the mayor's office is that it has permanently assigned two technical staff to support the farmers in their fields with technical advice, while also checking to guard against the use of agrochemicals. While the technician visits a member's farm, he is always joined by various other members of the group. This official control adds credibility to the participatory guarantee system, or PGS: peer control of organic production based on monitoring by the group members.

On Thursday, we visit the agroecological fair on Plaza 12 de Noviembre, which lasts from 8 AM to noon. All tables are covered with a bright orange cloth and the farmers all wear a green apron. While staff of the mayor's office provide all stands with biodegradable bags for consumers to carry their fresh produce, by 11 AM a band starts to play. Along with the permanent banners reminding shoppers of the agroecological fair every Thursday, this is an entertaining way to further draw in customers.

It is a lively event, and all the customers we talk to are pleased with this initiative that has been running for over a decade now. Local people say that there is a high rate of cancer and other diseases, which they attribute to the excessive use of pesticides on fruits and vegetables. The people visiting the market are fully convinced of the need to eat healthy food that is produced without chemicals. They are ready to pay a little extra, and they are pleased with the support of the mayor's office to make this happen.

The city government uses social media, TV, radio and print to further sensitize citizens to the need to eat healthy, local food, and to visit the various agroecological markets that are set up across the city. Consumers can also register at the weekly fair to take part in guided visits of the farms of some of the group members.

Helping consumers appreciate what it takes to produce healthy food and creating opportunities for consumers to interact with farmers directly, whether in the market or on their farms, cannot be left to farmers alone. Local authorities are indispensable for promoting healthy, local food cultures, as we have seen in this inspiring example of the city of Pelileo. As Mayor Maroto said in our video:

> Across the world, representatives of the cities and the towns should understand that it is the countryside that gives life to the city. So, one must understand that support for our farmers strengthens the local economy and that by strengthening the local economy, there is development and progress for the towns, and even more so if the products are organic.

This powerful statement from Pelileo's mayor shows how some leaders deeply appreciate the contribution of ecological farmers to their wider community. When this understanding is enshrined in policy, it can become permanent. In this case, after the video was published, Mr Maroto lost his bid for re-election, but his successor has made sure to keep supporting the agroecological markets.*

📹 Related video

Farmers, consumers and leaders from Pelileo, Ecuador, explain how to organize to produce and sell agroecological food in this video.

Agro-Insight and SWISSAID (2022) *Creating Agroecological Markets*. Hosted by Access Agriculture. 16 minutes.
www.accessagriculture.org/creating-agroecological-markets

---

*A version of the section above was previously published on www.agroinsight.com in 2022, by Paul Van Mele.

Marketing Something Nice

I've always been impressed by the way Bolivians adapt creatively to new situations. In 2018, my wife, Ana, and I went to a farmers' fair in the small town of Colcapirhua, near Cochabamba. The fair was supposed to be held in the charming main square of the town. Paved in flagstones, closed to car traffic and with steps leading up to a small church, it would have been a delightful venue. But some other local townspeople were already at the town square, loudly protesting about alleged corruption in their town council.

The protesters were there to stay, so the farmers moved their fair two blocks south, where they strung out their stands on an empty side lane along the main highway between Cochabamba and La Paz. It was less picturesque, but there were more potential customers passing by.

The farmers selling goods represented organized groups from all regions of Bolivia. The fair was actually part of the annual meeting of the National Soils Platform, which had chosen 'fair trade' as its annual theme. As we moved up the line of stalls, the farmers were keen to sell us a wide range of goods that were of high quality, but also unique, such as strawberries from the valleys of Santa Cruz, oven dried to sweet perfection.

Coffee growers from the Amazon (parts of which are cool enough for coffee) had brought little bags of coffee seed. 'Ready to plant!' they exclaimed, eager to encourage other farmers to start growing their own coffee. Cacao farmers from the Beni, also in the Amazon Basin, had bitter, white and milk chocolate. There was real pleasure in buying chocolate from the people who had made it from the cacao beans that they grew themselves.

There were tiny puffed grains of amaranth (ready to eat like cold cereal), fresh cherimoyas (*Annona cherimola*, a native fruit – but of a small, sweet variety that is now hard to find). Some farmers from Chuquisaca had a local variety of chilli that is so hot it is called *la gran putita* (the great little whore). We had to buy some.

There was traditional food, like an aged cow's cheese from the warm valley of Comarapa. It tasted marvellous, but the smell of cow was not for beginners.

What struck me the most was how many of the products were new, and inventive. Things you wouldn't find in the supermarket in Cochabamba, such as dried apples and preserved peaches still on the stone (moist and sweet but with no sugar added). Quinoa and wheat were packed in neat plastic bags, with labels, ready to make into soup.

As we say in the section 'Food for Outlaws' (see Chapter 13, this volume), smallholders with attractive products struggle to produce equally attractive labels, which by law often have to list ingredients. Here, the chocolate was wrapped in handsome paper with a printed label.

My favourite was the apple vinegar, in recycled Corona® beer bottles, made from clear glass. The farmers had covered the painted beer label with a new paper one, proudly explaining that this vintage was made from just three ingredients: organic apples, raw cane sugar with no additives, and water. The bottles were neatly sealed with bright yellow bottle caps.

Most of these farmers' associations have received support, often from Church-sponsored NGOs, some with volunteers from Europe and elsewhere. Outside help in manufacturing and packaging had clearly contributed to the quality of the goods, but the farmers were self-motivated to sell their goods. Agriculture is in large measure about producing something to sell.

Although this was an event on fair trade, there was no mention of being certified as Fairtrade. One speaker the first day had mentioned some of the hurdles that keep smallholders from being able to qualify for Fairtrade certification, and this group had readily agreed with her.

This group of smallholders certainly understood one basic idea, marketing means you must have something nice to sell: attractive, high-quality and well presented. Farmers across the globe deserve a fair price for their products, and smart marketing helps to achieve this.*

◼◼ Related video

Tomatoes can also be processed into juice, even in a West African village, as demonstrated in this video from Mali.

AMEDD (2017) *Tomato Concentrate and Juice.* Hosted by Access Agriculture. 12 minutes. www.accessagriculture.org/tomato-concentrate-and-juice

---

A version of the section above was previously published on www.agroinsight.com in 2018, by Jeff Bentley.

The Baker Farmers

The Organic World Congress only takes place once every 3 years, and in 2021 it was in France, which was lucky for me, because I had the pleasure of getting to know Olivier Clisson and his Irish wife, Lisa O'Beirne. On their inspiring biodynamic farm *Le Chant du Blé* (The Song of Wheat) on the outskirts of Rennes, they grow their own wheat and rye, various fruits, and they keep some farm animals. Neither Olivier nor Lisa grew up on a farm, but they have found their own passion for farming and for turning their harvest into delicious food.

Lisa always loved cooking, but her parents encouraged her to get a university degree, so she decided to study literature. Olivier moved from France to Belfast where he studied advanced maths and ancient Celtic history … and fell in love with Lisa. When they became young parents, in need of starting to earn money and not wanting to raise their son amidst the Northern Ireland conflict, they quit their studies and decided to move to Rennes, the capital of Brittany in France.

During their first years in France, they finished their studies. Lisa started working as a project manager in international business. It would take another 10 years before she was able to live out her dream to cook: for over a decade, Lisa ran the first registered organic restaurant in France. Olivier began to teach, in line with the tradition and expectations of his family, but when his dad passed away, Olivier took some time to reflect on his life. He decided he wanted to work with his hands in the earth and become a farmer. He embarked on a degree in biodynamic farming, the Professional Certificate in Agricultural Business Management (BPREA – *Brevet Professionnel Responsable d'Entreprise Agricole*), while working for 15 months as an intern on two different farms.

They bought an old farm house, started to renovate it, and began to farm. 'We didn't have a master plan. Things just happened', Lisa says.

One day Olivier attended a workshop on biodiversity where he met a traditional baker, who invited him over. 'When the baker opened the wood oven, the smell of fresh bread was so overwhelming that I realized that this is what I wanted to do. It was like a calling', says Olivier. He learned the art of sourdough breadmaking over the course of a year with a master baker.

At their farmhouse, Olivier opens a small room where he proudly shows his Astrié mill that he had tailor-made (Fig. 11.1). It allows for micrometric adjustment of the grinding stones, so that the cereal is no longer crushed between the millstones by the weight of the stone. This preserves the germ with all its nutrients.

> The only mills you find on the market are for large bakeries, and I needed something smaller to suit my needs. I can regulate the distance between the two granite millstones and grind my flour nearly constantly at a very low speed all day. My grains never get hot, so they are not precooked before making my breads and the flour keeps its full qualities.

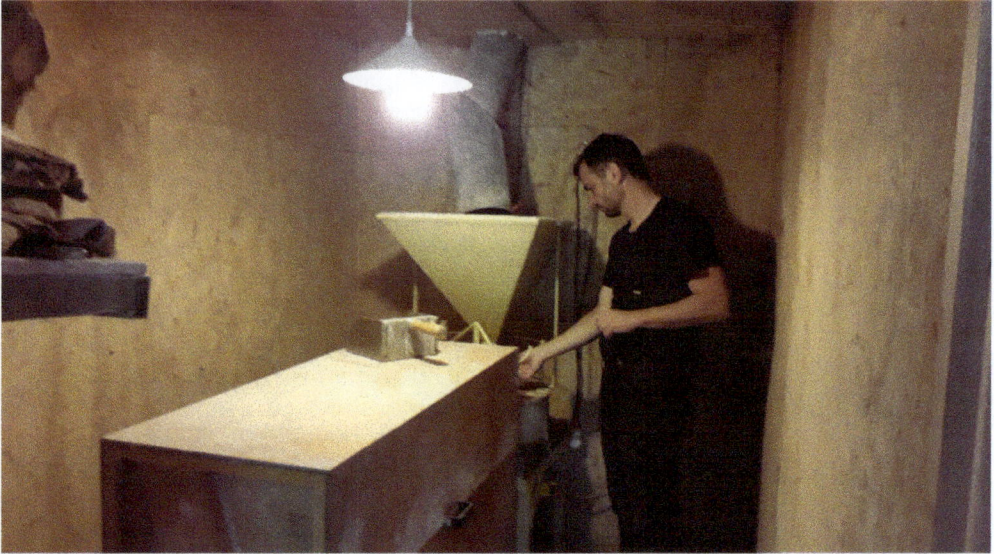

Fig. 11.1. Olivier Clisson, a biodynamic farmer in Brittany, France, grows his own wheat, which he grinds with a tailor-made mill. Photograph used with permission from Olivier Clisson.

Nothing is wasted: the bran is fed to four pigs that happily roam around a large, outdoor pen behind the farmhouse.

Olivier shows his small bakery room where he keeps his sourdough starter and the bread baskets woven from willow branches. Lined with a cotton cloth, the dough is left to rise in the baskets for 6h, after which he transfers the dough to the wood-fired oven. It is 5 PM, but when he opens the oven, just outside the baking room, some of the heat from the morning baking has remained. Olivier and Lisa had this oven built by an old man, a master of this unique skill, so it may be one of the last ovens built this way. The wood-fired oven is a major part of their unique approach to preparing food with deep respect for local knowledge (Fig. 11.2). Olivier confides:

> You have to be fully focused when making bread, as every day is different. The dough, the weather and also the wood we use to fuel the oven is different each day, so if you are tired and not fully mindful you will not make good bread that day.

Olivier bakes his bread at between 260 and 270°C, but there is no thermometer. After he removes the ashes and swipes the oven floor with a wet cotton cloth on a wooden stick, he measures the temperature using a method he invented.

According to Olivier, preparing bread is the easy part, but getting to master your wood oven has taken several years.

> I sprinkle some flour in the oven and count to four. If it turns black before I finish counting, the oven is still too hot and I sweep a few more times with the moist cotton cloth, or else my bread will be burned. If it takes longer

Fig. 11.2. Baked in a wood oven, the bread is pre-ordered and paid for in advance as part of food baskets compiled with ten other local organic farmers who produce everything from cheese to beef to pancakes. Photograph used with permission from Olivier Clisson.

than four counts, the oven is not hot enough and my bread will not be fully cooked.

Spread over 4 days a week, Olivier bakes about 250 kg of flour: mostly sourdough breads with a blend of 70% wheat and 30% rye. The 400 kg of bread are pre-ordered by two different CSA groups (community-supported agriculture): one from near their home village that has about 100 members, and one from the city of Rennes that has about 150 members. Olivier also bakes brioches and pies with a wide range of berries that they also grow on their farm. Lisa says:

> All that is baked is pre-ordered and paid in advance as part of food baskets that are prepared with about ten other local organic farmers who produce everything from cheese to beef to pancakes. Our bakery products are in such demand that whenever someone decides to stop their subscription and wants to get back in, they get on a waiting list and it may take 2 years before they can again order our bread.

Having just turned 50 years old in 2023, Olivier and Lisa have worked hard all their lives, but they are filled with joy and they are grateful to be able to do what they love to do, day after day.

Olivier also teaches and is now the head of a BPREA degree course on biodynamic farming. He is actively involved in the international federation of biodynamic farming, and has over the years hosted over 20 baker apprentices. 'Being able to pass on what we have learned to the younger generation is what has given us the most satisfaction', the couple concludes. And they have passed on their passion: their oldest son has decided to take over the farm, while their youngest son is continuing his studies in Paris to become a chef.

Besides the enormous popularity of their bakery products, Olivier and Lisa are driven by a rewarding quality of life and being able to constantly learn and explore. This inspiring couple shows how a family farm can produce honest food with respect for people and nature.*

◼️🎥 Related video

The story of *Le Chant du Blé* is captured in this video (in French, with subtitles).

Tous Terriens (2021) *Paysan-Boulanger/The Farmer-Baker*. Hosted by EcoAgTube. 10 minutes.
www.ecoagtube.org/content/paysan-boulanger-farmer-baker

---

*A version of the section above was previously published on www.agroinsight.com in 2021, by Paul Van Mele.

The Juice Mobile

Some 40 years ago, mobile services were common across Europe. In my home village, Kieldrecht in northern Belgium, I remember how a dairy farmer delivered bottles of fresh milk to his regular clients. The service was pretty sophisticated. The bottles came in different sizes, and several times a week customers ordered as many as they wanted, delivered on their doorsteps. Once a month people paid their bill. They would pay cash (after a short, friendly chat), or if they were away at work, they would leave the coins in an envelope on the doorstep for the milk man to collect. A quality, tailor-made service it was.

Early in the morning, two competing bakers delivered bread to their customers. My granddad, who lived by himself on his family farm, was happy to pay a few cents more for his bread, while receiving some friendly words from or sharing a joke with the baker's son.

But society has changed and most of those village services are long gone. Supermarkets and bread vending machines at the corner of streets have replaced the personalized services. Services have become more expensive, and today's customers no longer want to devote as much time to producing and processing food as our parents' and grandparents' generations did. But with an increased consciousness of healthy food and a desire from consumers to reconnect with food producers, some old services such as home deliveries have come back, while new initiatives such as farm shops and subscriptions to weekly vegetable boxes have become increasingly popular.

Last week I experienced a delightful local initiative to save some of our food and farming heritage in Limburg, the north-eastern province of Belgium. Limburg is known for its bicycle culture, greenery and orchards, all of which contribute to local tourism and the local economy.

Unfortunately, family orchards have suffered from the same hurried mentality of our generation; many orchards with tall, old varieties have been abandoned, with few new ones being planted.

To maintain the genetic diversity of old fruit varieties, the National Orchard Foundation established a genebank with over 3000 old and valuable local fruit varieties (van Laer, 2010). Besides their own orchards that serve as research, demonstration and training sites, the foundation has also successfully convinced people like my father-in-law to plant local fruit varieties.

Over the past 30 years, hundreds of people across Belgium have gradually re-established fruit gardens with old varieties. Until 2014, one could even get subsidies for planting tall fruit varieties. Still, my wife, Marcella, wondered:

> But how to keep young families motivated to plant and maintain fruit trees? Money to buy trees isn't the bottleneck, but time to maintain an orchard and process the fruit is. As fruit spoils fairly quickly, why would one bother to pick a few hundred kilograms of fruit?

That was until she learned about the Juice Mobile. On specific days throughout the fruit harvesting season (September to early November), the Juice Mobile comes to certain locations to process small batches of fruit from family orchards into fresh fruit juice.

The Juice Mobile is run by the National Orchard Foundation in collaboration with its sister organization, Nature and Orchards Social Workshop (NBSW – *Nationale Boomgaardenstichting Maatwerk*). During the harvesting season, the crew of three people works long days. On a day, they press up to 5 t of apples, pasteurize the juice and package it in 3- or 5-l sterile bags, neatly placed into cardboard boxes. The orchard owners go back home with all the juice from their fruit, having paid 1.25 Euros per litre for the service.

So, we called upon Marcella's sisters, brother, nephews and nieces. One day we picked some 600 kg of apples, and the next day brought them to the Juice Mobile. In an hour and a half, we had over 400 l of fresh juice.

The fruit pulp, along with the rotten and overripe apples that are sorted out at the beginning of the conveyer belt, is taken to a nearby farm to turn into biogas.

We spent two rewarding days: being out in the fresh air, picking fruit from the family's fruit trees, chatting with other people from the community. For a full year, the entire family will fondly recall that Marcella's dad planted those old apple varieties. Family ties and conservation of old fruit varieties can strengthen each other, with a little help from modern agro-processing.*

📽️ **Related video**

Women in Benin produce and market juice from fallen cashew apples, making the most of locally available resources.

DEDRAS (2016) *Preparing Cashew Apple Juice*. Hosted by Access Agriculture. 9 minutes. www.accessagriculture.org/preparing-cashew-apple-juice

---

A version of the section above was previously published on www.agroinsight.com in 2015, by Paul Van Mele.

Grocery Shops and Farm Shops

Few people realize how our food system is structured and how we consumers have a crucial influence. Exercising our food rights is as important as being politically active.

My dad ran a successful grocery store on the village market square, just across from the church. I still vividly remember the day when he took out an advertisement leaflet from the letter box. A year earlier a supermarket had opened in the village, accompanied by aggressive marketing. 'They sell the same orange juice cheaper than I can buy it from the wholesaler.' My dad turned to my mum, 'If this continues, I will have to close soon.' Customers from the neighbourhood suddenly started to avoid our shop as they walked to the supermarket, heads down, embarrassed because they no longer dared to greet my dad, with whom they had joked and chit-chatted for over 30 years.

Local entrepreneurs are resilient and creative. I am still amazed when I think of all the different goods my dad had on offer in his small shop, from fresh fruit to ice cream, from birdseed and toys to stockings for women. Along with my mum, he paid special attention to making the shop window as attractive as it could be during special occasions like Sinterklaas (6 December), Christmas and Easter. It was a real art that no supermarket could beat.

But shops need more than high-quality goods and services, and loyal customers. One day, the wholesaler who had sold produce to my dad for years bluntly announced that he could no longer supply us, as the wholesaler made more profit selling directly to the supermarkets and said it was not worthwhile continuing to supply independent retailers. By then, a second supermarket had already opened in the village. And so dad closed his shop. That was in the early 1990s. Dad was also a skilled printer, so he found other work. But he had loved his shop, because it let him make other people happy. Now that was gone.

Currently, in Belgium 95% of the food we eat is purchased from supermarkets, which continues to put small-scale, local entrepreneurs out of business. Supermarkets also harm local farmers by driving prices so low that farmers can barely cover their costs, as we described in the section 'Stuck in the Middle' (see Chapter 3, this volume).

Over the years, my wife, Marcella, and I have become good friends with Johan Hons and Vera Kuijpers, who grow organic vegetables and fruits and sell them in a farm shop they started about a decade ago. Each time we meet, they have some interesting stories to share. Vera said:

> We sell some of our produce to Biofresh, but they always pay the lowest possible price for our produce and prices have never gone up over the years.

I was already familiar with such practices that can really put the knife to farmers' throats, but had not expected this to happen in the organic food system, which I thought was fairer.

In 2019, Biofresh merged with the Dutch company Udea, after which economics started to overrule its philosophy. As Johan explained:

> Now Biofresh no longer allows retailers to enter its premises to see what fruit and vegetables are on offer if the amount they buy each week is below 1000 Euro, so many small farm shops like us have started to look for alternatives, but it is not easy.

Every Thursday, the day before their farm shop opens, Johan and Vera drive through half of Belgium to buy and sell fresh produce. Besides Biofresh, they now also buy from Sinature, BioVibe and directly from various farmer friends (see the section 'The Next Generation of Farmers' in Chapter 10, this volume).

Thirty years after my dad closed his village shop, the new farm shops which are providing fresh, healthy food are in the same stranglehold as the grocery shops in the 1990s. When profits overrule ethics, wholesalers decide who can buy from them and may cut off sales to small shops, just because the wholesaler wants even more money.

Small-scale retailers have higher transaction costs for stocking up, forcing many out of business. Supermarket chains are now buying up closed village shops to start specialty shops, and as irony would have it, to 'be closer to the customer'. Some supermarkets have even gone a step further, buying up organic farms and fishing grounds to gain full control over the food we eat. Supervised by managers, the real farmers and fisher folks with a passion for their profession risk becoming mere employees devoid of any decision-making power.

The European Green Deal provides an action plan to boost the efficient use of resources by moving to a circular economy, restoring biodiversity and cutting pollution (European Commission, 2019). Yet it remains to be seen what measures will be put in place to support our small-scale farmers, farm shops and community initiatives such as weekly boxes of fresh local produce procured through group purchasing associations.

Without appropriate measures, organic farming risks becoming a variation of industrial agriculture with emerging opportunities captured by a few dominant food chain actors, who further consolidate their power, wealth and decision-making over what food we get on our table.

In the meantime, we consumers should not underestimate our influence. As Johan said, 'consumers have the market in their hands.' Buy local from farm shops, farmers' markets and small-scale retailers as much as you can. The supermarkets' claim that they are 'local' is misleading and pushes those with a passion for their profession out of business.*

📹 Related video

The demand for organic products is also rising in Africa, where farmers are learning to sell in this market.

FIBL (2021) *Marketing Organic Products in Africa*. Hosted by EcoAgTube. 58 minutes. www.ecoagtube.org/content/marketing-organic-products-africa

---

A version of the section above was previously published on www.agroinsight.com in 2020, by Paul Van Mele.

Strawberry Fields Once Again

Like many Bolivians, Diego Ramírez never thought about remaining in the village where he was born and starting a business on his family's small farm. As a kid, he loved picking fruit in his grandparents' strawberry patch in the village of Ucuchi, and swimming with his friends in a pond fed with spring water, but he had to leave home at a young age to attend high school in the small city of Sacaba, and then he went on to study computer science at the Public University of San Simón (UMSS – *Universidad Mayor de San Simón*) in the big city of Cochabamba, where he found work after graduation.

Years later, Diego's dad called his seven children together to tell them that he was selling their grandparents' farm. It made sense. The grandparents had died, and the land had been idle for about 15 years. Yet, it struck Diego as a tragedy, so he said, 'I'll farm it.' Some people thought he was joking. In Ucuchi, people were leaving agriculture, not getting into it. Many had migrated to Bolivia's eastern lowlands or to foreign countries, so many of the fields in Ucuchi were abandoned. It was not the sort of place that people like Diego normally return to.

When Diego decided to revive his family farm in 2018, he turned to the internet for inspiration. Although strawberries have been grown for many years in Ucuchi, and they are a profitable crop around Cochabamba, Diego learned of a commercial strawberry farm in Santo Domingo, Santiago, in neighbouring Chile, that gave advice and sold plants. Santo Domingo is 2450 km from Cochabamba, but Diego was so serious about strawberries that he went there over a weekend and brought back 500 strawberry plants. Crucially, he also learned about new technologies like drip irrigation, and planting in raised beds covered with plastic sheeting. Encouraged by his new knowledge, he found dealers in Cochabamba who sold drip irrigation equipment and installed it, along with plastic mulch, a common method in modern strawberry production.

Diego was inclined towards producing strawberries agroecologically, so he contacted the AGRECOL Andes Foundation, which was then organizing an association of ecological farmers in Sacaba, the small city where Diego lives (half way between the farm and the big city of Cochabamba). In that way, Diego became a certified ecological farmer under the Participatory Guarantee System, Agroecological Farmers of Sacaba (SPG PAS – *Sistema Participativo de Garantía, Productores Agroecológicos Sacaba*). Diego learned to make his own biol (a fermented solution of cow dung). Now he mixes biol into the drip irrigation tank, fertilizing the strawberries one drop at a time.

Diego also makes his own organic sprays, like sulfur-lime brew and Bordeaux mix. He applies these solutions every 2 weeks to control powdery mildew, a common fungal disease, thrips (a small insect pest), red mites and damping off. I was impressed. A lot of people talk about organic sprays, but few make their own. 'It's not that hard', Diego shrugged, when I asked him where he found the time.

Diego finds the time to do a lot of admirable things. He has a natural flair for marketing and has designed his own packing boxes of thin cardboard, which he had printed in La Paz. His customers receive their fruit in a handsome box, rather than in a plastic bag, where fruit is easily damaged. He sells direct to customers who come to his farm, and at agroecological fairs and in stores that sell ecological products.

Diego still does his day job in the city, while also being active in community politics in Ucuchi. He also tends a small field of potatoes and is planting fruit trees and prickly pear on the rocky slopes above his strawberry field.

Diego has also started a farmers' association with his neighbours, ten men and ten women, including mature adults and young people who are still in university. The association members grow various crops, not just strawberries. Diego is teaching them to grow strawberries organically and to use drip irrigation. To encourage people to use these methods, he has created his own demonstration plots. He has divided his grandparents' strawberry field into three areas: one with his modern system, one with local varieties grown the old way on bare soil, with flood irrigation, and a third part with modern varieties grown the old way. The modern varieties do poorly when grown the way that Diego's grandparents used. And Diego says the old way is too much work, mainly because of the weeding, irrigation, pests and diseases.

Ucuchi is an attractive village in the hills, with electricity, running water, a primary school and a small hospital. It is just off the main highway between Cochabamba and Santa Cruz, an hour from the city of Cochabamba where you can buy or sell almost anything. Partly because of these advantages, some young people are returning to Ucuchi. Organic strawberries are hard to grow, and rare in Bolivia. But a unique product, like organic strawberries, and inspired leadership can help to stem the flow of migration, while showing that there are ways for young people to start a viable business in the countryside. Diego clearly loves being back in his home village, stopping his pickup truck to chat with people passing by on the village lanes. He also brings his own family to the farm on weekends, where he has put a new tile roof on his grandparents' old adobe farm house.

Agriculture is more than making a profit. It is also about family history, selling a high-quality product locally, being a good neighbour and finding work that is satisfying and creative.*

◼ Related video

AGRECOL Andes promotes a participatory guarantee system (PGS) to offer farmers a way to organize, and self-guarantee that their quality produce is organic.

Agro-Insight and AGRECOL Andes (2023) *A Participatory Guarantee System.* Hosted by Access Agriculture. 11 minutes.
www.accessagriculture.org/participatory-guarantee-system

A version of the section above was previously published on www.agroinsight.com in 2020, by Jeff Bentley.

Your Own Piece of Land

In the sections above, we have written about the value of growing and processing one's own food. For people who don't own their own land, one alternative is the allotment.

Allotments or community gardens are small plots cultivated by individuals who abide by rules set by the landowner, often a local council, but sometimes a church, a private company or individual willing to provide a social service. Non-commercial gardeners pay a modest annual rent against the security over a longer-term land tenancy.

While a new trend of urban gardening is sparked by a young generation in favour of eating healthy food that is produced with minimal food miles, few people realize that allotment schemes originated out of a need for food security.

Across most of Europe, industrialization in the 19th and early 20th century drove people from the countryside to the cities in search of jobs. Their working and living conditions were often appalling and, coupled with poor nutrition, meant that early deaths in a family were common (Nicholas and Oxley, 1993). Church authorities and local councils started 'gardens of the poor'. Railroad companies also allotted plots of land to their workers. The stretches of land along the sides of the railway were unsuitable for general agriculture, but offered a good opportunity for the large workforce to grow their own food. Through this social service, companies kept their workforce happy.

During the First and Second World Wars it became a real challenge to bring enough food from the countryside to the cities. Most of the male workforce was called up by the armed forces. Fuel was also rationed and prioritized for moving soldiers, weapons and supplies. As ships were no longer able to import as much food (Holland, 2010), the British government launched a 'digging for victory' campaign that used waste ground, railway edges, gardens, sports fields and golf courses for farming or vegetable growing. Victory gardens were also planted in back yards and on apartment-building rooftops. By 1943, the number of allotments had peaked at an estimated 1.75 million (Matless, 2016).

To support newcomer growers, many of whom did not have prior farming expertise, numerous radio and TV programmes were developed to strengthen people's skills while at the same time instilling a communal pride in the nation (Matless, 2016).

When looking at today's allotment plots, a few things strike the eye: first, each plot shows a unique mix of innovations as tenants experiment to get the best out of their garden. And second, the soil is often quite black, indicating the many years the soil has been nourished with organic matter. Long-term leases encourage gardeners to cherish the land and invest in its future.

Throughout history and across countries, allotment gardens have taken many shapes and forms. Many that were started during the hardships of war continued long into peacetime, in part because of demands from gardeners who loved being outdoors and growing their own produce. While allotments often started as poverty relief, they now help salaried professionals unwind from the stress of the office. Like agriculture, gardening evolves, and does more than just produce food.*

◼◼ Related video

Nematodes are a problem in many small gardens. But they can be managed ecologically.

Agro-Insight (2016) *Managing Vegetable Nematodes*. Hosted by Access Agriculture. 16 minutes.
www.accessagriculture.org/managing-vegetable-nematodes

A version of the section above was previously published on www.agroinsight.com in 2017, by Paul Van Mele.

References

European Commission (2019) The European Green Deal. Available at: https://ec.europa.eu/info/strategy/priorities-2019-2024/european-green-deal_en (accessed 16 October 2024).

Holland, J. (2010) *The Battle of Britain: Five Months that Changed History; May–October 1940*. St. Martin's Griffin, New York.

Matless, D. (2016) *Landscape and Englishness*. Reaction Books, London.

Nicholas, S. and Oxley, D. (1993) The living standards of women during the industrial revolution, 1795–1820. *The Economic History Review* 46, 723–749.

van Laer, P. (2010) Importance of diversification in the variety assortment of apple, pear, cherry and plum orchards. *Mitteilungen Klosterneuburg* 60, 458–462.

12 Reviving Traditional Food Cultures

Abstract

Traditional food is among the best in the world, and it can be brought to life with some creative community engagement. For example the NGO Nawaya uses a kitchen training centre in Egypt to help rural women become empowered caterers, by preserving and sharing their culinary heritage. A school in Peru integrates local knowledge and biodiversity into its curriculum, with students learning through hands-on activities like establishing a school garden. *La Tablée*, a festive event in France, brings together organic farmers and consumers, fostering appreciation for local food traditions. In Bolivia, the NGO AGRECOL Andes empowers smallholder farmers to sell affordable organic produce in low-income communities. These cases underscore the potential of community engagement to revive traditional food cultures for a wider audience.

The Kitchen Training Centre

During the annual Access Agriculture staff meeting in 2022 in Cairo, Egypt, we learned about a creative way to build rural women's skills and confidence, while reviving traditional food culture.

One afternoon, our local colleague Laura Tabet, who co-founded the NGO Nawaya about a decade ago, invited us all to visit Nawaya's Kitchen Training Centre. None of us had a clue as to what to expect. Walking through the gate, we were in for one surprise after the next.

Various trees and shrubs border the green grass on which a long table is installed. Additional shade is provided by a porch of woven reeds from nearby wetlands. The table is covered with earthenware pots containing a rich diversity of dishes, all new to us. But before the feast starts, we are invited to have a look at the kitchen.

Hadeer Ahmed Ali, a warmly smiling staff member of Nawaya, guides the 20 visitors from Access Agriculture into the spacious kitchen in the building at the back end of the garden. Several rural women are quickly moving small earthen pots in and out of the oven, while others add the last touches to some fresh salads with cucumber and parsley. The kitchen, with its stainless steel and tiled working space, is immaculate, and the dozen women all wear identical yellow aprons. Their group spirit is clear to see.

We are all separated from the cooking area by a long counter. When Hadeer translates our questions into Arabic, the rural women respond with great enthusiasm. One is holding a camera and takes photos of us while we interact with her colleagues. It is hard to imagine that some of these women had never left their village until a year and a half ago, when Nawaya started its Kitchen Training Centre.

Later on, Laura tells me that each woman is from a different village and is specialized in a specific dish:

> We want each of them to develop their own product line, without having to deal with competition from within their own village. While traditional recipes are the basis, we also innovate by experimenting with new ingredients and flavours to appeal to urban consumers.

The women source from local farmers who grow organic food, and cater for various events and groups. In the near future, they also want to grow some of their own vegetables and sell their own branded products to local shops and restaurants, and even deliver to Cairo.

The women have sharpened their communication skills by regularly interacting with groups of schoolchildren from Cairo. But becoming confident to interact with foreigners and tourists from all over the world is a different thing. Hence Nawaya engaged Rasha Fam, who studied tourism and runs her own business. She taught the women how to interact with tourists. Unfortunately, it is against Egyptian law to bring foreign tourists to places like this; tour operators can only take tourists to places that are on the official list of tourist destinations. The tourism industry in Egypt is a strictly regulated business.

Rasha also confirms what we had seen: these rural women are genuine and, when given the opportunity, it brings out the best in them. The training programme helped women calculate costs, standardize recipes, host guests and deliver hands-on activities in the farm and the kitchen.

When we walk out of the Kitchen Training Centre, a few women are baking fresh *baladi* bread (traditional Egyptian flatbread) in a large gas oven in the garden (Fig. 12.1). Large wooden trays display the dough balls on a thin layer of flour. One of the ladies skilfully inserts her fingers under a ball to transfer it to a slated paddle-shaped tool made from palm fronds (locally called *mathraha*). When she slightly throws the flat balls up, she gives the *mathraha* a small turn to the left. With each movement the ball becomes flatter and flatter until it is the right size. With a decisive movement she then transfers the flatbreads into the oven.

Nandini, our youngest colleague from India, is excited to give it a try. Soon also Vinjeru from Malawi and Salahuddin from Bangladesh line up to get this experience. We all have a good laugh when we see how our colleagues struggle to do what they just observed. It is a good reminder that something that may look easy can in fact be difficult when doing it the first time, and that perfection comes with practice.

Our appetite raised, we all take our places around the table, vegetarians on one side. The dishes reveal such a great diversity of food (Fig. 12.2). It

Fig. 12.1. Rural Egyptian women are trained at Nawaya's Kitchen Training Centre. By reviving traditional food culture, they gain confidence and establish their own food business. Photograph used with subjects' permission.

Fig. 12.2. Diverse dishes offered to groups of visitors help to raise consumer awareness about healthy, traditional food.

is not every day that one has a chance to eat buffalo and camel meat, so tender that they surprise many of us. The vegetarians are delighted with green wheat and fresh pea stews.

Traditional food cultures can be promoted in many different ways. What we learned from Nawaya is that when done in an interactive way, it helps to build bridges between generations and cultures. People are unique among vertebrates in that we share food (Flandrin and Montanari, 1999). Eating and cooking together can be a fun, cross-cultural experience.

Whether the diners come from Cairo or from distant countries, they love to interact with these rural women to experience what it takes to prepare real Egyptian food. Nawaya's Kitchen Training Centre has clearly found the right ingredients to boost people's awareness about healthy, local food cultures.*

▣ Related video

Egyptians have made and eaten pressed dates since before the times of the Pharaohs. Dates are still part of a modern economy, grown and processed on small farms.

Nawaya (2016) *Making Pressed Dates*. Hosted by Access Agriculture. 12 minutes. www.accessagriculture.org/making-pressed-dates

---

A version of the section above was previously published on www.agroinsight.com in 2022, by Paul Van Mele.

Agroecology for Kids

The Third of May is a special day for the people of Tres de Mayo, Huayllacán, in Huánuco in the highlands of central Peru. On that day they celebrate the community's anniversary.

When Jeff, Marcella and I reached the village at 7:30 AM, half an hour before the start of school, local women in their beautiful traditional dresses and hats with colourful flowers were already frying trout in a big pan over a wood fire.

Soon after, breakfast was served in one of the classrooms of the local school. We took our seats on the small chairs and we were soon joined by the teachers. Large pots with steaming food were carried into the classroom in a wheelbarrow. We all enjoyed the trout with a diversity of boiled potatoes, four or five different varieties on each plate. And diversity would be the theme of the day, as the school had organized a fair on agrobiodiversity and local food.

After raising the flag of Peru and singing the national anthem, children from all classes performed a variety of songs, poems and sketches, all related to their rich farming, food and herbal health traditions. The school walls were lined with the drawings of the pupils from kindergarten to secondary school. Later on, a jury evaluated the children's artwork. When I asked Dante Flores, one of the jury members, how they make a selection, he said that they would all get a prize. This is a nice way to ensure all kids feel appreciated.

The school had a specially designed classroom, which they call the Seed Room or *Muru Huasi* in the local Quechua language. The name is a little confusing as it is not a seed storage room, but rather it sows the seeds of respect and love for local culture, agrobiodiversity and ecological farming.

Besides a library with books on Peru and its natural richness, the room also displays drawings and models made by the schoolchildren that show the different plants and animals their parents have on their farms. Various games, such as an adaptation of Monopoly®, trigger the children to learn in a joyful way about their rich farming and food traditions. On one of the tables, a variety of clay-made tubers are on display, in all sorts of shapes and colours that I have never seen before.

Outside, at the fair, I realize that these kids have not improvised, but that Andean tubers indeed are incredibly diverse. For potatoes alone, of which Peru is the centre of origin, the country has some 4000 varieties (see the section 'Native Potatoes' in Chapter 5, this volume).

The fair is also visited by parents, and several are *conservacionistas*. Jeff and I talk to one of them. Eustaquio Hilario Ponciano tells us that he and his family have about 300 potato varieties that he mixes all together in his fields. This strategy offers diverse food, but also spreads risk as one is never sure what the weather will be like.

A day earlier, we went with some joyful, singing children up to the high country at about 4000 m altitude to meet don Eustaquio and his wife as they harvested one of their potato fields. From the steep slopes they unearthed red, yellow, black and purple spotted potatoes in all shapes. The children repeat in a loud voice the Quechua names of each variety. It is clear how much the children enjoy being out in the field and listening to some of the stories behind the names of the local varieties.

At the fair, long tables displayed a selection of the potato varieties along with other Andean tubers, such as oca (*Oxalis tuberosa*), olluco (*Ullucus tuberosus*) and mashua (*Tropaeolum tuberosum*).

From the children's drawings, songs and poems, it was clear how much pride the school director, Luz Valverde, and the teachers take in their rich culture, and how they have successfully passed this on to the children.

We hope that the video that we made on teaching ecological farm and food culture in schools will inspire educators around the world. Children deserve healthy, nutritious food that is grown locally, free of chemicals.*

◼️ Related video

Below is a link to the video we filmed in Tres de Mayo.

Agro-Insight, CIZA and IDMA (2022) *Teaching Agroecology in Schools.* Hosted by Access Agriculture. 14 minutes.
www.accessagriculture.org/teaching-agroecology-schools

A version of the section above was previously published on www.agroinsight.com in 2022, by Paul Van Mele.

The School Garden

Learning by doing is one of the most powerful educational approaches, both for adults and for children. Kids observe what their parents and others in the community do, and they copy it.

For a week, Marcella, Jeff and I, accompanied by our local partners Aldo Cruz and Dante Flores, were eating breakfast in the house of doña Zenaida Ramos and her husband, who are also members of the parents' committee of the Tres de Mayo school in Huayllacayán, in central Peru.

As we walked to the school, which combines a kindergarten, primary and secondary school, we could see that activities on the playground had already started. Today was the final day of our filming trip in this community, and the school had planned the installation of a school garden: an activity they used to have before COVID-19 and which they are now more than happy to resume.

When we met the dynamic school director, Luz Valverde, she told us:

> Here in the district of Huayllacayán more than 600 varieties of native potatoes are grown. And we have so many varieties of oca, olluca and mashwa, as well as plenty of medicinal plants. We also have broad beans and barley. So, the students should value all that we have here in our community. And they should also know how much nutritional value these products have, so they can keep conserving them. They should conserve our environment, conserve the local biodiversity that exists here.

Some of the teachers and pupils of the final year of primary school had started loosening the hard soil and breaking the clods with picks, and we decided to give them a helping hand. It was hard work and, at an altitude of 3000 m, we often had to pause to gasp for breath. Some 15 minutes later, various mothers and fathers had also joined. We immediately saw how experienced they were doing this type of work. The kids could see it too. The boy next to me carefully observed how one of the dads was handling the tool and then soon resumed with regained insight and energy.

Some other farmers arrived with their Andean foot plough, or *chaquitaclla* in the local language Quechua (Morlon, 1992). This pre-Inca tool is still widely used and considered the best tool for preparing land on steep rocky slopes without causing soil erosion. When I saw the tool in use on the flat playground of the school, I was amazed how fast it loosened up big clods of soil.

Within an hour we had prepared the garden plot, stretching 25 m long and 3 m wide. As we worked the last few metres, one of the fathers stopped me and tried to make it clear to me, in Spanish, that I had to work the land differently here. I hadn't noticed, but the tail end of the plot slightly slants downward, so we needed to make the planting ridges perpendicular to the other ones, so that the irrigation water could easily move and cover the entire plot. This shows how farmers think ahead in everything they do. While I was just preparing the soil, he was already thinking about irrigation.

After being shown how to plant, the kids soon started to plant onions, lettuce, broccoli and cabbage seedlings. The parents assisted, observing their kids, at times giving them some advice. Later on, they would add some of the native crops.

By 10 o'clock the work was over. At one side of the school garden, the kids installed a netted fence to protect the food garden when they play on the playground. As they fixed the long green net with pieces of metal wire to wooden poles, previously inserted in the soil by the teachers, it was great to see how handy these kids are: it is clearly not the first time that they have used all these tools.

School gardens have been around for many years and in many countries, as a way for students to learn about farming and to provide healthy ingredients for school meals, free of agrochemicals. What is unique here is that many parent farmers helped to install the garden, so the kids learn by observing their parents in a school context. Perhaps most important is that the kids feel that the culture of their parents is being appreciated by a formal institution. All too often children are taught to look down on their own indigenous culture and on local customs and knowledge.

'I think that school directors across the world who work as leaders should include everyone in our educational community, so that everything we have around us is valued', concludes Luz Valverde in her interview.*

◼️ Related video

Baby food is not often appreciated as part of local food culture, but infant porridge is the first step, and it can be made entirely from local ingredients.

AMEDD (2016) *Enriching Porridge*. Hosted by Access Agriculture. 13 minutes.
www.accessagriculture.org/enriching-porridge

A version of the section above was previously published on www.agroinsight.com in 2022, by Paul Van Mele.

La Tablée

The choice to eat healthy, organic food cannot be left to consumers alone. While organizing farm visits to inform and build trust among consumers is important, too often such initiatives are left to individual farmers. But when this is coordinated at a higher level with multiple stakeholders, including local authorities, an amazing dynamism can be created, as I learned in 2021 during a visit to France.

With colleagues from Access Agriculture, we decided to stay a few days longer in Rennes, after we had attended the Organic World Congress in September 2021. Strolling through the historic city centre towards the old church of Saint George, we were pleasantly surprised to discover *La Tablée* (Table Guests), a festive open-air event on the grounds around the ruins where people are invited to taste local products laid out on long lines of picnic tables.

After Rennes city applied to host the Organic World Congress, a group involved with the application went on to create *La Tablée* and various other events. They called their group 'Voyage to Organic Lands'.

After some friendly volunteers explained the concept, we took a seat and started to taste some of the apple juices, which were all delicious and remarkably distinct. Each bottle has a name printed on the bottle screw cap (Arthur, Lancelot, Merlin, Gauvain, Vivianne, Perceval and Excalibur). Before France was unified in 843 AD, Britain (*la Grande Bretagne*) and Brittany (*la Petite Bretagne*) had close ties, and historians increasingly believe that the legend of the hero king Arthur and his brave knights has its roots in France, in the forests near Rennes (Warren, 2000). Perhaps French apple juice or cider was served at the round table.

When I heard someone speaking about apples over the loudspeakers, I realized that there was a live radio show being broadcast from one of the corners. Radio Rennes was interviewing the organic apple grower Arnaud Lebrun. Openly and honestly, Arnaud explained how he started his career as a salesman for a pesticide company. Arnaud explains live on air:

> After more than a decade, I began to see all the damage this was doing to the environment, and I could no longer find peace with myself. I decided to quit my job and make a 180-degree shift. My wife and I bought a neglected apple orchard with trees that were already 40 years old and we converted it into an organic apple orchard. We had to learn everything. I did not even know how to drive a tractor.

In the shade of an old oak tree, interviews went on all day long with local farmers and food producers. While we only stayed on for an hour or so, I could still hear Arnaud's wife profess, 'Our customers truly appreciate all the products we make from our apples. What gives me the most satisfaction is to see the smiles on people's faces.'

Brittany has the richest diversity of apple varieties in France, and a long tradition of producing cider and *pommée*, a thick sweet preserve to spread on bread. Preparing the *pommée* is a community event that celebrates harvest, as the women clean the apples while men take turns all night long stirring the thickening *pommée* in a huge copper pot over a fire.

Another remarkable traditional product on the picnic tables is *gwell*, a creamy type of yoghurt made by fermenting raw milk from the Bretonne Pie Noir, a breed of local cow that almost went extinct in the 1970s. *Gwell* is traditionally eaten with flat round buckwheat cakes (*galette*) or potatoes, and is an excellent ingredient for desserts.

As we are having a great culinary experience, Lisa and Olivier, the sympathetic local baker farmers whom we just got to know at the Organic World Congress, arrive and join our table (see the section 'The Baker Farmers' in Chapter 11, this volume). They brought with them some more fresh bread and other traditional goodies.

Small leaflets, each one with a little quiz, invite people to reflect on one particular aspect of making and eating food. This pleasant event brings consumers and producers closer to each other, and with the radio reaches a much wider audience.

For over 60 years, consumers have been influenced by marketeers to eat and drink overprocessed foods, stripped of their nutrients (Pollan, 2008). It will take time for people to switch from flavour-enhanced junk to real food. Through joint efforts between organic and biodynamic farmer associations, researchers and restaurant owners, as well as authorities from cities and regions, changing consumer behaviour towards healthy, natural food can become a continuous, concerted effort.

As I learned that week in Rennes around the table, consumers and farmers need more than connections, they need to form communities, and a bit of fun can help.*

◤■◣ Related video

It's not *pommée*, but in Africa traditional porridge is enriched with juice from the baobab fruit.

Hochschule Rhein-Waal and Biovision (2020) *Enriching Porridge with Baobab Juice.* Hosted by Access Agriculture. 8 minutes.
www.accessagriculture.org/enriching-porridge-baobab-juice

A version of the section above was previously published on www.agroinsight.com in 2021, by Paul Van Mele.

Naturally Affordable

Around the world, certified organic farmers often complain that they need higher prices for their produce, but this means that they will only sell to people with money. The poor won't have access to this healthy food. I learned this recently from Mariana Alem, a Bolivian biologist of the AGRECOL Andes Foundation, which is working with smallholder producers to grow and sell affordable organic food in low- and middle-income areas in the Cochabamba Valley in Bolivia.

Since 2019, Mariana and her colleague María Omonte, an agronomist, have worked with 36 farmers, mostly women, who had already organized themselves to sell produce in local fairs. The farmers self-declared that their produce was free of agrochemicals. To build rapport in the group, the women organized themselves to visit each other for a peer review. It started as a kind of inspection, but as the women got to know each other, these visits became a chance to exchange seeds or to share information about topics like recipes for controlling pests without chemicals.

Mariana and María found one group of these farmers at a market called El Playón, in a low-income neighbourhood on the edge of the urban sprawl of metropolitan Cochabamba. At this market, buyers and sellers were dressed in work clothes, wearing the broad-brimmed hats of rural women. They were speaking Quechua, as country people do, rather than Spanish, as is spoken in the city.

Since the market only started in 2019, it still had an unfinished look. The stalls were handmade from rough lumber.

Paul and Marcella and I met doña Gladys, who was selling tomatoes for 8 Bolivianos (US$1.15) per kilogram, a competitive price. Most of the other women sold their *locotos* (hot peppers) or cucumbers in small piles for 5 Bolivianos each, units that poor people were used to buying.

Others were selling cut flowers and fruit. One of the older women, doña Saturnina, was selling organic peaches for the same price as conventional ones. Doña Saturnina, who is joined at her stall by her granddaughters, also gave us a glass of juice made from fresh peaches boiled in water – so refreshing.

To offer organic produce at affordable prices, one trick is for farmers to sell directly to consumers. This way, the farmer can charge the retail price. It is easier said than done, because selling is work.

Mariana and María had mentored the women, helping them to make aprons and to identify themselves as self-declared ecological producers. They were not formally certified, but they were producing without agrochemicals.

To sell to retail customers, you have to be at the fair on every market day with a good diversity of products. This group plants many different crops, rather than each person planting the same thing. Then, by sitting near each other, the women can attract and share customers. They also increase their range of products on offer by reselling some things they buy from their farmer neighbours and from wholesalers. Unfortunately, that means that not all of their produce is organic.

With the wisdom of hindsight, Mariana and María insist that the main thing is to be transparent with the consumers, when farmers adopt these strategies to supply their stall. When the women come to sell, each one has a green cloth where they pile up their organic produce. They also have an orange cloth where they are supposed to display conventional vegetables. The distinction has been a bit too subtle for consumers and a little too hard for the farmer-sellers to manage.

'Did you forget your orange cloth today', some vendors will chide the others, who are selling conventional produce from their green blanket.

These are the kinds of learning experiences that one may have while setting up something new. These growing pains aside, the point is that selling healthy, organic food should be for everyone, not just for people who can afford to pay extra. It is important for farmers to get a fair price, while making organic food affordable. Healthy eating shouldn't be a luxury.*

📽 **Related video**

The video we filmed with AGRECOL Andes, about how to sell ecological food at local fairs, is available in English, Spanish, Quechua, Aymara and French.

Agro-Insight and AGRECOL Andes (2023) *How to Sell Ecological Food*. Hosted by Access Agriculture. 15 minutes.
www.accessagriculture.org/how-sell-ecological-food

A version of the section above was previously published on www.agroinsight.com in 2023, by Jeff Bentley.

References

Flandrin, J.-L. and Montanari, M. (1999) *Food: A Culinary History from Antiquity to the Present.* Columbia University Press, New York.

Morlon, P. (1992) *Comprendre l'Agriculture Paysanne dans les Andes Centrales: Pérou - Bolivie (Understanding Peasant Agriculture in the Central Andes: Peru - Bolivia)*. Institute de La Recherche Agronomique (INRA), Paris, France.

Pollan, M. (2008) *In Defense of Food: An Eater's Manifesto*. Penguin Books, London.

Warren, M.R. (2000) *History on the Edge: Excalibur and the Borders of Britain, 1100–1300*. University of Minnesota Press, Minneapolis, Minnesota.

13 Organic Certification, Regulations and Policy

Abstract

Well-intentioned regulations often need adjustments to support a diverse and sustainable food system, as shown through several case studies. Food safety has evolved from combating deliberate adulteration (like brick dust in chilli powder) to navigating vulnerable globalized food chains. Piglets can get the iron they need by rooting in the soil instead of taking supplements, but Belgian regulations make that a controversial solution. Although EU regulations allow mobile slaughterhouses for better animal welfare, Belgian policies favour large-scale systems, making it harder for smallholders and local food processors to operate. A Bolivian law intended to protect consumers with food labelling may unintentionally put small-scale producers out of business. In modern Europe, hunting is regulated to control pests (like wild boars and pigeons) and maintain wildlife balance. New EU regulations on seed coatings (like the neonicotinoid ban) have increased the problem of pigeons for conventional farmers, but hunters fail to provide their services in a monocrop landscape stripped of hedgerows.

Eating Bricks

In Belgium we have an expression: 'all Belgians are born with a brick in their stomach', meaning that everyone aspires to build their own house someday. But when bricks are literally eaten, something has gone seriously wrong.

Some 25 years ago, during one of my first projects in Sri Lanka, news came out that chilli powder was being mixed with ground-up bricks. Some crooks were trying to make a dishonest profit. Ground chilli and powdered bricks are of a similar colour and consistency. Few buyers taste the chilli powder when they buy it, and as chilli is typically added to sauces, never eaten straight, a cheating dealer, supplying to regional or international markets for customers he would never see, might get away with such a scam.

© Jeffery W. Bentley and Paul Van Mele 2025. *Agroecology in Practice: From Local Initiatives to Global Scaling Through Video* (J.W. Bentley and P. Van Mele) DOI: 10.1070/0781800628793.0013

Fortunately, in Europe we have a long history of food safety standards, regulations and government institutes safeguarding the quality of the food that enters the market and ends up on our plates. But such systems are absent, dysfunctional or just getting started in many developing countries.

Yet many developing countries have an advantage when it comes to food safety: short food chains. Control measures on food safety are less important when one relies on short food chains. In Sri Lanka, for instance, I used to patronize spice gardens where urban people would stock up on black pepper, chilli or cardamom. Over the years the customers would establish a relationship based on trust with the family running the spice garden. Even in the markets, most vendors know their regular customers and would never risk selling them a fake product. Suppliers are motivated to sell high-quality products to their valuable, steady customers.

I had forgotten about this incidence of adulterated chilli until recently. While reading the book *The True History of Chocolate* (Coe and Coe, 1996), I was struck by one particular paragraph on food adulteration. Cacao had spread from Latin America to Portuguese, Spanish, English and French colonies across Africa and Asia in the 19th century. In 1828, the Dutch chemist Coenraad Van Houten took out a patent on a process to make powdered chocolate with a very low-fat content. The Industrial Revolution was in full swing and entrepreneurs in England and America established their first companies to make chocolate for the masses. For centuries chocolate had only been known as a foamy drink, consumed mainly by royalty, the aristocracy and clergy.

Already in 1850, the British medical journal *The Lancet* mentioned the creation of a health commission for the analysis of foods. According to the journal, suspicions about the quality of the mass-produced chocolate proved correct: in 39 out of 70 samples, chocolate had been adulterated with red brick powder. Similar results were obtained from samples of chocolate seized in France. The investigations led to the establishment of the British Food and Drug Act of 1860 and the Adulteration of Food Act of 1872 (Coe and Coe, 1996).

A similar trend took place in the milk industry.

In Belgium, starting in 1900, machines were deployed to scale up butter production. Just 2 years later, the Belgian farmers' organization, the *Boerenbond* (Farmers' League), decided to employ food consultants to check the administration, hygiene and quality of the dairies. In 1908, the *Boerenbond* established a food laboratory which it deemed necessary to help curb the increase in butter adulteration (Belgische Boerenbond, 1990).

Now, more than a century later, the COVID-19 pandemic has exposed once more the vulnerability of a globalized food system with long supply chains. Slightly more than 50% of all food produced in Belgium is exported, including milk. As the demand from China dropped, this left farmers unable to sell dairy, meat and potatoes. Belgian dairy cooperatives, like others in the global dairy sector, also struggled to secure sufficient packaging material, as this relied on imports of certain materials (Acosta *et al.*, 2021).

Such troubles are triggering people to rethink how to make our food system more sustainable. For a long time, food safety regulations were assumed to be the main pillar of a safe food system, but the pandemic has revealed that the complexity of a global food system makes it prone to breaking down, leaving producers and consumers vulnerable. Over the years, overly rigid food safety standards in Belgium have discouraged farmers from adding value to their own produce and selling it on their farms. More will hopefully be done in the near future to encourage farmers to process and sell food on their farms. In these short food chains, farmers will be motivated to make clean, healthy products.

Food in Europe is reasonably safe and healthy, but COVID-19 has shown us how modern food systems are fragile. Burdensome regulations oppress smallholders until they are not even able to make a cheese for their neighbours. By investing in shorter food chains, we can make our food systems more resilient and bring back the distinctive flavours of local foods. Shorter, more adaptable food chains will build trust, while leaving the bricks to those who are building houses.*

◼️🎥 Related video

How to make high-quality chilli powder, without bricks.

Agro-Insight (2016) *Making Chilli Powder*. Hosted by Access Agriculture. 11 minutes. www.accessagriculture.org/making-chilli-powder

---

A version of the section above was previously published on www.agroinsight.com in 2021, by Paul Van Mele.

Iron for Organic Pigs

Organic agriculture is steadily on the rise across the globe (Willer *et al.*, 2021). As the sector grows and more farmers convert from conventional to organic farming, policies and regulations are being fine-tuned. Many conventional farmers took out big loans to invest in modern pig houses. Balancing animal welfare with a heavy debt burden is a delicate exercise, as I recently learned from my friend Johan Hons, a long-time organic farmer in north-eastern Belgium, who said:

> When some 40 years ago a neighbour farmer offered to let me use one of his vacant stables, I bought my first Piétrain pigs (a Belgian breed of pig) and started rearing. In those early years, I always supplemented iron. A few years later, Vera and I were able to start our own farm. We were convinced that organic farming was the only way food should be produced, so I gave my pigs the space to roam around in the field. Ever since then, they never needed any iron injections and they never got sick.

Iron is an essential mineral for all livestock. Iron-deficient piglets will suffer from anaemia: they will remain pale, stunted, have chronic diarrhoea and, if left untreated, they will die. Worldwide, piglets in conventional agriculture are commonly injected with a 200-mg dose of iron a few days after birth (Merlot, 2022). Although this intramuscular injection is effective against anaemia, it is very stressful to the piglets, and the long suckling periods in organic pig farming can still lead to iron deficiency (Heidbüchel *et al.*, 2019).

In a natural environment a rooting sow acquires enough iron from the soil, which she passes on to the suckling piglets through her milk. But most pigs in conventional farming in Belgium are raised on slatted floors and have no access to soil. Sows only have enough iron reserves for their first litter. Piglets of the second and third litter would already have a shortage of iron and become sick, unless given supplements.

Under Belgian regulations, organic meat pigs are allowed only one medical treatment for whatever illness. If a second treatment is given, pigs can only be sold in the conventional circuit and hence farmers do not get a premium price. With more conventional farmers eager to convert to organic to earn a higher income, members of Bioforum, the Belgian multi-stakeholder platform for organic agriculture, asked the regulatory authorities whether iron injections could be considered as a non-medical treatment.

As a member of Bioforum, Johan suggested an alternative: 'When the sow delivers in the pigsty, I daily give her piglets a few handfuls of soil from the moment they are 1 week old. I put it out of reach of the sow, otherwise she would eat it, and continue doing so until the piglets are a few weeks old and allowed outside. Just like human babies, the piglets have a curious nature and by giving them early access to soil, they immediately build up their iron stores and immunity.'

For Johan, caring for animals is knowing what they need and providing for their needs and comfort throughout their life. This starts at birth.

However, his suggestion initially got a cold reception at the forum, whose members also include retailers. Most farmers who want to convert to organic cannot let their pigs roam on the land, because on conventional Belgian farms, concrete is more abundant around the pig houses than soil.

And however creative they found Johan's suggestion to provide piglets with soil in the stables, this was not considered a feasible option. Conventional farmers have invested heavily in modern pig houses with slatted floors and automated manure removal systems, and bringing in soil would obstruct the system. Adjusting such houses to cater for organic farming is an expense few farmers are willing to make.

Belgian authorities decided that, because of a lack of commercial alternatives to iron injections, they would be temporarily accepted in organic agriculture, on the condition that the iron formulation is not mixed with antibiotics.

A sustainable food system is at the heart of the European Green Deal. As the European Commission (2023) has set a target under its Farm to Fork Strategy to have 25% of the land under organic agriculture by 2030, it will need to reflect on how far the regulations for organic agriculture can be adjusted, and above all on possible measures to support farmers to convert.

Regulations can force expensive changes, such as large stables for livestock, and may drive smallholders out of business, while denying the pigs the freedom to happily root around in the earth. If left to the pigs to decide, they would surely opt for more time outdoors and less concrete around their houses, not a tweak in the regulations to declassify iron injection as a medical treatment.*

◼◼ Related video

Farmers in Uganda show us a practical, low-cost way to house pigs, which allows them to source more of their own iron.

Environmental Alert and Farmers Media (2020) *Housing for Pigs*. Hosted by Access Agriculture. 13 minutes.
www.accessagriculture.org/housing-pigs

---

*A version of the section above was previously published on www.agroinsight.com in 2021, by Paul Van Mele.

Mobile Slaughterhouses

An article on the BBC News (2019) reminded me of how policy makers can look at narrow technical solutions (how to kill an animal) while ignoring broader, yet largely unquestioned issues about how we organize our food system. I will illustrate this by giving an example of my former neighbour, René, a farmer who lives in the east of Belgium.

René inherited the farm from his father. EU subsidies in the 1980s encouraged farmers to increase the number of livestock, so by the time his father handed over the farm there were around 1000 pigs. But René of course had to pay his brothers for their share of the inheritance. By the time he was in his early 50s he was still paying off loans to the bank. With the low price he got from selling to supermarkets, René realized he had to find a way to earn more money. He decided to take a butchery course and soon after he started selling meat products directly to the public on his farm.

By 2010, René had reduced his herd to some 200 pigs. Now, he still sells some pigs to supermarkets, but his main income is derived from selling meat from his own animals to people who visit his farm butchery. Every Monday morning René takes two pigs to the slaughterhouse, spends the week processing the meat into more than 20 products ranging from salamis to smoked hams and pâtés, and then he and his wife, Marij, open the shop from Friday to Sunday.

With a great sense of pride, René told me a few years back that he had finally paid off all his debts. But just a year later, the farm family had to take another major decision. The nearest slaughterhouse in Genk, some 20 km from his farm, had closed down, so René was forced to drive over 50 km to have his animals slaughtered.

Regulations required that for longer distances, live and slaughtered animals had to be transported in special vehicles. René told me this would cost the family around 10,000 Euros, not counting the extra distance to be travelled each week. One has to sell a lot of sausages to pay for this extra cost. Closing the farm and going to work in a factory was not an option, so they kept their heads high, invested in a trailer and the family continued with their farm and food business.

It seemed that the slaughterhouse in Genk that René relied on had closed down under pressure from certain lobby groups in favour of a food system in which supermarkets influence the rules. More public debate is needed in Belgium on how policies can support smallholder farmers and local food initiatives rather than discourage them.

EU legislation allows mobile slaughter of all kinds of domestic animals. Under the supervision of the farmer and the professional slaughterer who drives the mobile abattoir, animals can be spared the stress of long transport and be slaughtered humanely at home (Hultgren *et al.*, 2020). Belgian politicians can learn from countries where such initiatives are in use, such as those in Scandinavia, France, Australia and New Zealand.

Food is power, and a democratic food system is one that is owned and controlled by as many people as possible, instead of by a few giant companies (Global Alliance for the Future of Food, 2021). While community-supported agriculture can give people a sense of ownership over their food, more is required to fundamentally change our food system with due respect given to the people who produce the bulk of our food: professional and passionate smallholder farmers. Mobile abattoirs deserve more attention to enhance the welfare of animals and to keep farmers crafting food in a business they are proud to run.*

◼◖ Related video

Farmers can also make sausages from rabbits.

KENAFF, FRT Malawi, AIS Egypt and Malawi Polytechnic (2020) *Making Sausages from Rabbit Meat*. Hosted by Access Agriculture. 12 minutes.
www.accessagriculture.org/making-sausages-rabbit-meat

A version of the section above was previously published on www.agroinsight.com in 2019, by Paul Van Mele.

Food for Outlaws

A law can have unintended consequences, as I learned at the national meeting of *Prosumidores* (producers + consumers) held in Cochabamba, Bolivia. Their second annual meeting, in 2017, promoted healthy, local food and family farming, bringing together farmers and concerned consumers. It was held in a grand old house in the city centre. Half a dozen groups of organized farmers sat at tables in the entrance way, selling fresh chillies, local red apples, amaranth cookies, delicious wholewheat bread and little flasks of apple vinegar, among other unusual and wonderful products. A few had labels, but none had a list of their ingredients or nutritional qualities.

When the presentations started in the main room, most of the farmers stayed outside where potential customers were still looking at the goods. The farmers had come to sell, not to hear lectures.

Inside the large hall, one of the talks was by a government lawyer. She gave a helpful explanation of Law 453, on consumers' food rights, signed in 2013. And while it had been the law of the land for 4 years, many consumers were unaware of it. Law 453 is a complex piece of legislation which aims to promote safe and healthy food and includes interesting bits such as 'promoting education about responsible and sustainable consumption.' But the lawyer caught the most attention when she explained that the law required all food to have a label, listing the ingredients and the nutritional characteristics of the food.

That is when a perceptive woman from the audience rose to make a statement.

> I'm opening a shop to sell agroecological foods, but if I adhere strictly to this law, I won't be able to buy products from the kinds of people who are selling just outside this door.

There was a moment of stunned silence, because it was true. Few smallholders can design and print a label listing the nutritional qualities of their products. (For example, I bought some fresh, delicious wholewheat bread at the meeting. Many people could write a list of ingredients in a homemade product like bread, but would not know how to list the calories or other nutritional qualities of the food.)

The more food is regulated, the more difficult it will be for small producers to meet well-meaning standards. At this event, the lawyer was unable to answer the storekeeper's question. It seemed as if no one had noticed the potential legal difficulties for smallholders (even organized ones) to sell packaged food.

This law was written to keep consumers safe, and it was certainly never intended to prevent smallholders from selling their produce directly to consumers; organized peasant farmers are a key constituency of the current government. The anti-smallholder bias was simply an unintended consequence of the law, a bit of thoughtlessness.

That this well-intentioned law has never been enforced is a blessing in disguise, because in Bolivia many people (especially low-income women) still sell food from their homes, on street corners and in open-air markets. In just the first few weeks of 2024 I have had fresh cheese, lucerne (alfalfa) sprouts and roasted cacao beans: all excellent and none quite in line with the laws for food labelling. Over-regulating this market would make it harder for some people to make a living, while eliminating some of the more diverse and interesting foods in Bolivia.*

◢ Related video

Smallholders around the world make snacks at home, to sell and to add value to their produce.

Agbangla, A. (2018) *Making Groundnut Oil and Snacks*. Hosted by Access Agriculture. 12 minutes.
www.accessagriculture.org/making-groundnut-oil-and-snacks

A version of the section above was previously published on www.agroinsight.com in 2017, by Jeff Bentley.

The Village Hunter

From time to time I run into our village hunter, Pol Gielen, which is always a good occasion to get to know the village history a little better, and to learn about the changing challenges of hunters and farmers alike. In our village, Erpekom in north-eastern Belgium, with only 300-odd citizens, Pol Gielen is one of the two people allowed to hunt on the village grounds. The licence has been passed on from generation to generation. While hunting in Europe is a centuries-old occupation, it has not always had the same social relevance.

William the Conqueror, the Norman King who reigned in England from 1066 until his death in 1087, introduced the French tradition of Royal Forests – land that was marked out as royal hunting grounds. A decade earlier, William allied himself with Flanders, now part of Belgium, by marrying Matilda, daughter of Count Baldwin of Flanders. William was a fervent hunter who loved being in the woods. A peasant caught hunting could be thrown into prison or, just as likely, publicly executed. For centuries to follow, hunting became a stylized pastime of the aristocracy (Davies, 2023).

In contemporary Europe, hunting is no longer confined to the rich. While hunting licences are to ensure that only well-trained persons are allowed to hunt, the right to hunt is also linked to the duty to care for all animals listed in the hunting laws. For various species, such as deer, wild boars, hares and pheasants, hunters and authorities have to develop plans, detailing how many animals may or must be killed during the hunting season. Some pest species, such as pigeons, can be shot with little restriction.

In the section 'Bullets and Birds' (see Chapter 8, this volume), I wrote how factories coat seed with chemicals to repel pigeons. However, organic farmers refuse to use that seed, preferring to call on local hunters to come to the rescue if need be. My recent encounter with Pol, our village hunter, showed me how changing pesticide regulations in Europe continue to influence the relationships between hunters, farmers and the environment.

In 2018, the European Commission banned three neonicotinoids (synthetic nicotinoids, toxins originally derived from tobacco). The ban covers all field crops, because these pesticides harm domesticated honey bees and wild pollinators (Stokstad, 2018). Neonics, as they are commonly called, are often coated onto seeds to protect them from soil pests. These pesticides are systemic, meaning they spread through the plant's tissue. The toxin eventually reaches pollen and nectar, where it harms pollinators. According to a study by Professor Dave Goulson in the UK, most seeds and flowers marketed as 'bee-friendly' at garden centres, supermarkets and DIY centres, like Aldi and Homebase, are contaminated with systemic pesticides (Goulson, 2017; Lentola *et al.*, 2017; see also Malone, 2018). In fact, 70% of the plants contained neonics, commonly including the ones banned for use on flowering crops by the EU (Goulson, 2017). Birds, bees, butterflies, bats and mammals are indiscriminately poisoned when they forage on contaminated plants.

The dramatic decline of bees and other pollinators due to the use of neonics and other pesticides is threatening the sustainability of the global food supply. Of the 100 crop species that provide 90% of global food, 71 are pollinated by bees (Stokstad, 2018).

To further reduce the negative impact of agriculture on the environment, more restrictions have been imposed because of mounting evidence that pesticide-coated seed is also harmful to birds, including partridges, a favourite game bird for 1000 years that has now become a rarity. Apart from providing subsidies for installing and maintaining hedgerows around farmers' fields to serve as food and nesting habitat for birds, the European Commission recently banned methiocarb, a toxic insecticide used as a bird repellent, often used to coat maize seed (European Commission, 2019).

With the new EU regulations limiting seed coatings, conventional dairy farmers got worried that birds would damage their maize crop and have begun looking for alternatives. That is the reason why one of our farmer neighbours decided to call upon Pol, the village hunter. It was when Pol was on his way back from that farmer that I ran into him and he said, 'Well, the farmer asked me to come and shoot pigeons, but I told him: "I would be happy to help you, but where do you want me to hide, you have removed all the hedges in your fields!"'

Regulations to curb the indiscriminate and dangerous use of pesticides on seed and in fields must go hand in hand with other measures, such as promoting hedgerows that fulfil important ecological functions for birds and pollinators. Also, environmentally friendly alternatives could be further investigated and promoted. Green, innovative technologies, such as clay coating, are likely to become increasingly important. Clay is perceived by insects and birds as soil and offers a natural protection for the seeds. The clay can even be enriched with other natural additives to repel birds and insects.

Fortunately, regulations have not killed this local tradition of hunting. No longer the pastime of kings, hunting can be part of a regulated, rural policy that allows people to manage bird pests, without the use of chemicals, while saving the bees.*

📹 **Related video**

Further ideas on managing birds, not always by killing them.

Agro-Insight and CIAT (2019) *Managing Birds in Climbing Beans*. Hosted by Access Agriculture. 10 minutes.
www.accessagriculture.org/managing-birds-climbing-beans

A version of the section above was previously published on www.agroinsight.com in 2020, by Paul Van Mele.

References

Acosta, A., McCorriston, S., Nicolli, F., Venturelli, F., Wickramasinghe, U. *et al.* (2021) Immediate effects of COVID-19 on the global dairy sector. *Agricultural Systems* 192, 103177.

BBC News (2019) Research into benefits of mobile abattoirs. Available at: www.bbc.com/news/uk-scotland-highlands-islands-46958906 (accessed 16 October 2024).

Belgische Boerenbond (1990) *100 Jaar Boerenbond in Beeld. 1890–1990 (An Overview of 100 Years of Boerenbond in Pictures. 1890–1990)*. Dir. Eco-BB – S. Minten, Leuven, Belgium.

Coe, S.D. and Coe, M.D. (1996) *The True History of Chocolate*. Thames and Hudson, London.

Davies, R. (2023) The Charter of the Forest: Your guide to the 13th-century law. *History Extra* (official website for BBC History Magazine and BBC History Revealed). Available at: www.historyextra.com/period/plantagenet/charter-forest-what-why-important/ (accessed 16 October 2024).

European Commission (2019) The European Green Deal. Available at: https://ec.europa.eu/info/strategy/priorities-2019-2024/european-green-deal_en (accessed 16 October 2024).

European Commission (2023) From Farm to Fork: Our Food, Our Health, Our Planet, Our Future. Available at: https://ec.europa.eu/commission/presscorner/detail/en/fs_20_908 (accessed 16 October 2024).

Global Alliance for the Future of Food (2021) *The Politics of Knowledge: Understanding the Evidence for Agroecology, Regenerative Approaches, and Indigenous Foodways*. Global Alliance for the Future of Food. Available at: https://futureoffood.org/insights/the-politics-of-knowledge-compendium/ (accessed 16 October 2024).

Goulson, D. (2017) Pesticides in "Bee-Friendly" flowers. Available at: www.sussex.ac.uk/lifesci/goulsonlab/blog/bee-friendly-flowers (accessed 16 October 2024).

Heidbüchel, K., Raabe, J., Baldinger, L. and Hagmüller, W. (2019) One iron injection is not enough – Iron status and growth of suckling piglets on an organic farm. *Animals* 9, 651.

Hultgren, J., Segerkvist, K.A., Berg, C., Karlsson, A.H. and Algers, B. (2020) Animal handling and stress-related behaviour at mobile slaughter of cattle. *Preventive Veterinary Medicine* 177, 104959.

Lentola, A., David, A., Abdul-Sada, A., Tapparo, A., Goulson, D. *et al.* (2017) Ornamental plants on sale to the public are a significant source of pesticide residues with implications for the health of pollinating insects. *Environmental Pollution* 228, 297–304.

Malone, K. (2018) Beeware! "Bee-Friendly" Garden Plants Can Contain Bee-Harming Chemicals. Available at: www.bumblebeeconservation.org/beeware-bee-friendly-garden -plants-can-contain-bee-harming-chemicals/ (accessed 16 October 2024).

Merlot, E. (2022) Improved health, welfare and viability in young pigs: Oral iron supply in neo-natal piglets to avoid anaemia. POWER-Factsheet, no. 2.4. Research Institute of Organic Agriculture FiBL, Frick, Switzerland. Available at: https://orgprints.org/id/eprint/ 43622/1/13_POWER_Piglets_Ironsupply_einzeln_Web.pdf (accessed 16 October 2024).

Stokstad, E. (2018) European Union expands ban of three neonicotinoid pesticides. *Science*. Available at: www.science.org/content/article/european-union-expands-ban-three-neon icotinoid-pesticides (accessed 16 October 2024).

Willer, H., Trávníček, J., Meier, C. and Schlatter, B. (2021) *The World of Organic Agriculture – Statistics and Emerging Trends 2021*. Research Institute of Organic Agriculture FiBL and IFOAM – Organics International, Frick, Switzerland and Bonn, Germany.

14 Money Matters

Abstract

Financial viability in agroecology presents both challenges and opportunities. Youths in Africa, for instance, face hurdles in accessing land, capital and training, despite their interest in agriculture. Similarly, Bolivian women transitioning from conventional flower production to organic vegetables encounter difficulties securing fair prices and managing weekly deliveries of personalized food baskets to urban consumers. Complex cost–benefit analyses are often required, as exemplified by a Bolivian project integrating apple trees with vegetables. However, success stories like a small-scale watermelon farmer in the Solomon Islands and a large-scale roundtable event in Bolivia, where dozens of farmers' associations had a chance to interact with domestic and international traders, demonstrate diverse pathways to financial prosperity in agroecology. The experience of a company producing organic fertilizers highlights the potential for both smallholders (who can produce their own inputs) and larger farms (who can buy them) to improve soil health, emphasizing the value of both commercial and on-farm solutions in replacing chemical inputs.

Youth Don't Hate Agriculture

Rural youth are moving to the cities by the busload. Yet counter to the prevailing stereotype, many young people like village life and would be happy to go into farming, if it paid. This is one of the insights from a study of youth aspirations in East Africa that unfolds in three excellent country studies written by teams of social scientists, each working in their own country. Each study followed a parallel method, with dozens of interviews with individuals and groups in the local languages, making findings easy to compare across borders.

In Ethiopia many young people grow small plots of vegetables for sale, and would be glad to produce grains, legumes, eggs or dairy. Youth are often attracted to enterprises based on high-value produce that can be grown on the small plots of land that young people have (Endris and Hassan, 2020).

© Jeffery W. Bentley and Paul Van Mele 2025. *Agroecology in Practice: From Local Initiatives to Global Scaling Through Video* (J.W. Bentley and P. Van Mele)
DOI: 10.1079/9781800628793.0014

Youth are also eager to get into post-harvest processing, transportation and marketing of farm produce, but they lack the contacts or the know-how to get started. Ethiopian youth have little money to invest in farm businesses, so they often migrate to Saudi Arabia where well-paid manual work is available (or at least it was, before the pandemic).

In northern Uganda, many youths wanted to get an education and a good job, but unwanted pregnancies and early marriage forced many to drop out of secondary school. If dreams of moving to the city and becoming a doctor, a lawyer or a teacher don't work out, then agriculture is the fallback option for many young people. But, as in Ethiopia, young Ugandan farmers would like their work to pay more (Boonabaana *et al.*, 2020).

In Tanzania, many youths have been able to finish secondary school and some attend university. Even there, young people go to the city to escape poverty, not to get away from the village. Many youths are even returning, like one young man who quit his job as a shop assistant in town to go home and buy a plot of land to grow vegetables. Using the business skills he learned in the shop, he was also able to sell fish, and eventually invested in a successful 2-ha cashew farm (Mwaseba *et al.*, 2020).

These three insightful studies from East Africa lament that extension services often ignore youth. But the studies also suggest to me that some of the brightest youth will still manage to find their way into agriculture. Every urban migrant becomes a new consumer, who has to buy food. As tropical cities mushroom, demand will grow for farm produce.

If youth want to stay in farming, they should be able to do so, but they will need better access to land, capital and training in topics like natural ways to keep soil, plants and animals healthy, as well as food processing and how to make their produce more appealing for urban consumers. Improved infrastructure will make country life more attractive, and more productive. Better mobile phone and internet connectivity will help link smallholders with buyers and suppliers. Fresh food will reach the cities faster over good roads. A constant electricity supply will allow food to be processed, labelled and packaged in the countryside. New information services, including online videos, can also help give information that young farmers need to produce high-value produce.*

🎥 Related video

Young people in Kenya have developed a business as pest control scouts. The video is available in more than 30 languages, having inspired youth across the Global South to help farmers tackle the devastating fall armyworm.

Agro-Insight and FAO (2018) *Scouting for Fall Armyworms*. Hosted by Access Agriculture. 14 minutes.
www.accessagriculture.org/scouting-fall-armyworms

*A version of the section above was previously published on www.agroinsight.com in 2021, by Jeff Bentley.

The Struggle to Sell Healthy Food

Consumers are increasingly realizing the need to eat healthy food, produced without agrochemicals, but on our recent trip to Bolivia we were reminded once more that many organic farmers struggle to sell their produce at a fair price.

In early 2023 Jeff, Marcella and I were filming with a group of agro-ecological farmers in Cochabamba, a city with 1.4 million inhabitants at an altitude of over 2500 m. Traditionally, local demand for flowers was high, to use at weddings, funerals and other gatherings, but that all ended with COVID-19. Vegetables could still be sold, but farmers needed training to produce them ecologically. When we interview Nelly Camacho on camera, she explains why the women were motivated to give up chemical-intensive farming: 'We contaminate the environment, we contaminate our Mother Earth and we contaminate our health.'

As the women began to produce vegetables instead of flowers, they also took training on ecological farming. They realized that the only way to remain in good health is to care for the health of their soil and the food they consume. All of them being born farmers, the step to start growing organic food seemed a logical one. With the support of AGRECOL Andes, an NGO that supports agroecological food systems, a group of 16 women embarked on a new journey, full of new challenges.

'Over these past years, we have seen our soil improve again, earthworms and other soil creatures have come back. But I think it will take 10 years before the soil will have fully recovered from the intense misuse of flower growing', says Nelly.

On Friday morning, we visited the house of one of the members of the group. Various women arrived, carrying their produce in woven bags on their backs. Their fresh produce was harvested the day before, washed, weighed, packed and labelled with their group certificate. Internationally recognized organic certification is costly, and most farmers in developing countries cannot afford it. So, they use an alternative, more local certification scheme, called the Participatory Guarantee System or PGS, whereby member producers evaluate each other. More recently, the group has also received certification from the national government agency the National Service for Agricultural and Livestock Health and Food Safety (SENASAG – *Servicio Nacional de Sanidad Agropecuaria e Inocuidad Alimentaria*).

AGRECOL staff support the women as they prepare food baskets for their growing number of customers who want their food delivered either to their home or to their office. Some customers also come and collect their weekly basket at the AGRECOL office. Jeff's wife, Ana, shows us one evening how every week she receives a list of about four pages with all the produce available that week, and the prices. Until noon on Wednesday, the 150 clients are free to select if and what they want to buy. The orders are passed on to the farmers, who harvest on Thursday, and the fresh food is delivered on Friday morning: a really short food chain with food that has only been harvested the day before it is delivered.

Organizing personalized food baskets weekly is time-consuming. Most farmers also need institutional support as they lack a social network of potential clients in urban centres. AGRECOL has invested a lot in sensitizing consumers about the need to consume healthy food, using leaflets, social media, fairs and farm visits for consumers. Without support from AGRECOL or someone who takes it up as a full-time business, it is difficult for farmers to sell their high-quality produce.

In her interview, Nelly explains that the home delivery was a recent innovation they introduced when the COVID-19 crisis hit, as local markets had closed down, yet people still needed food. Now that public markets have reopened, demand strongly fluctuates from one week to the next, and with the tight profit margins, it might be a challenge to turn it into a profitable business. NGOs like AGRECOL play a crucial role in helping farmers produce healthy food and raising the awareness of consumers, who learn to appreciate organic produce.

As Cochabamba is a large city, AGRECOL has over the years helped agroecological farmer groups to negotiate with the local authorities to ensure they have a dedicated space on the weekly markets in various parts of the city.

Local authorities have a crucial role to play in supporting ecological and organic farmers that goes way beyond providing training and inspecting fields. Farmers need a fair price and a steady market to sell their produce. Being given a space at conventional urban markets and dedicated agroecological markets is helping, but in low-income countries few consumers are willing to pay a little extra for food that is produced free of chemicals. Public procurements by local authorities to provide schools with healthy food may provide a more stable source of revenue. It is no surprise that global movements such as the Global Alliance for Organic Districts (GAOD, 2023) have made this a central theme.

Agroecological farmers who go the extra mile to nurture the health of our planet and the people who live on it deserve a stable, fair income and peace of mind.

As Nelly concluded in her interview:

> So, for that reason we need to become educated, and most of all educate the children, the youth.*

◼️📷 Related video

AGRECOL Andes also promotes a participatory guarantee system (PGS), used in many countries to offer farmers a way to organize and self-guarantee that their quality produce is organic.

Agro-Insight and AGRECOL Andes (2023) *A Participatory Guarantee System*. Hosted by Access Agriculture. 11 minutes.
www.accessagriculture.org/participatory-guarantee-system

A version of the section above was previously published on www.agroinsight.com in 2023, by Paul Van Mele.

What Counts in Agroecology

Measuring the costs and benefits of a small farm can be harder than on a large one, especially if the small farm includes an orchard and makes many of its own inputs, as I saw on a visit in 2019 to Sipe Sipe, near Cochabamba, Bolivia, where a faith-based organization, *Agroecología y Fe* (Agroecology and Faith), is setting up ecological orchards.

The director of Agroecology and Faith, Germán Vargas, explained that a forest creates soil, gradually building up rich, black earth under the trees, while agriculture usually exposes the soil to erosion. A farm based on trees, with organic fertilizer and with vegetables growing beneath the trees, should be a way to make a profit while conserving the soil. Extensionist Marcelina Alarcón showed us the apple trees that she and local farmers planted in August 2018. They started by terracing the 1 ha of gently sloping land.

Next, they built an irrigation tank. In 1 week of hard work, they built a 200,000-l circular water reservoir of stone and concrete (gravity-fed with stream water) to irrigate the terraces and three additional hectares. The cost was 64,000 Bolivianos (US$9,275), which seems like a big investment, but similar reservoirs built 30 years ago are still working.

Lush beds of lettuce, cabbage, broccoli, wheat and onions were thriving beneath the apple trees. When one crop is harvested another takes its place, in complex rotations over small spaces. No chemicals are used, but the group makes calcium sulfate spray and liquid organic fertilizers to improve the soil, prevent crop diseases and enhance the production and quality of the apples and vegetables.

The group has harvested vegetables four times and sold them directly to consumers at fairs organized by Agroecology and Faith for a total gross receipt of 4,380 Bolivianos (US$635).

I was visiting the farm at Sipe Sipe with a small group organized by Agroecology and Faith and some of their allies. Some of the lettuce, onions and tomatoes from the farm end up in a tub during our visit, to make a salad for the visitors – part of a fabulous lunch (complete with fresh potatoes and mutton cooked underground) offered at a modest price. Produce cooked on site and sold informally on the farm is probably not counted when estimating profitability. After the tour of the farm and before the lunch, Marcelina set up a table with some vegetables for sale. She was kept quite busy writing down each transaction as we bought small bags of tomatoes and other produce for less than a dollar each.

The sale of half a kilogram of tomatoes is as much work to document as the sale of 20 t of rice. A small farm has many more sales than a large one and it takes a lot of administrative work to keep track of produce that is not sold because it is harvested for seed, feed or for the family table.

The cost–benefit of a conventional field is simpler to tabulate: so much labour, machinery, seed and chemicals, all purchased, and single crop yields measured with relative ease. Yet this doesn't tell the whole story. Loss of soil due to erosion, or carbon and nitrogen released into the atmosphere, or pollution from fertilizer runoff all have a cost, even if they are often dismissed as 'externalities' (Macháč *et al.*, 2021).

An agroforestry system, like the hectare of apples and vegetables we visited, starts with a large investment in irrigation and terracing. Many of the inputs are labour, or homemade fertilizers, and their cost is not always counted. The apple trees have not yet borne fruit, and some of the vegetables may escape the bookkeeper's tally. Yet here the 'externalities' have a positive and valuable contribution: soil is being created, carbon is being sequestered, chemical pollution is nil and livelihoods are enriched as local farmers, mostly women, learn to work together to produce healthy food to sell. Classical economic comparisons with conventional farms fail to take account of these benefits.

A recent report from the FAO (the UN's Food and Agriculture Organization) concludes that yield data is too poor a parameter to compare conventional (over-ploughed, chemical-intensive) agriculture with agroecology, a beyond-organic agriculture with soil conservation and respect for local communities (HLPE, 2019).

Even a small farm can be complicated and returns hard to estimate. Until we learn to measure the environmental efficiency as well as financial profitability of agroforestry or agroecological farms properly, they will never look as good on paper as they really are.*

◼ Related video

This video discusses finances on a small farm.

Philippine Permaculture Association, Alangilan National High School and NISARD (2024) *Record-Keeping for Integrated Farming*. 15 minutes.
www.accessagriculture.org/record-keeping-integrated-farming

A version of the section above was previously published on www.agroinsight.com in 2019, by Jeff Bentley.

Other People's Money

In the Solomon Islands, Osanti Ludawane shows that it is possible to make money from a small farm. Osanti grew up in Takwa, near the north end of the island of Malaita. He graduated from high school in Honiara (the capital city) and then took 2 years of accounting school. After graduation, Osanti worked as an accountant before he 'got tired of counting other people's money', as he put it. So, he took a 3-month course on a Taiwanese vegetable development farm, on growing fruits and vegetables, and then came back to Takwa to grow watermelons.

He harvests the fruit during the Christmas and New Year season and takes a few hundred melons by truck and boat to Honiara. He sells them and brings back chicken manure, but that costs a lot of money. So Osanti knows that the next step is to lower fertilizer costs.

When I met Osanti in September 2013, a lot of his neighbours were following his example, planting watermelons as a cash crop and still growing taro, sweet potatoes and cassava to eat at home.

Osanti used many ingenious innovations. He showed us the little cages of sticks that people pound into the earth around the crab holes in the watermelon fields. These 'land crabs', widespread throughout the Indo-Pacific, dig tunnels 2 m deep. The tunnel ends in a burrow, which has more than one exit. These highly territorial crabs live in the same burrow their whole life. When the tunnel entrances are disturbed or blocked, the crabs usually repair them within a week, and always as close as possible to the original entrance (Foale, 1999; Hurley, 2012). This is enough to give the watermelon plants a respite, and the new crab door will be predictably close to the old one, and the farmers can easily peg it shut again.

Paul and I wondered what would happen if everyone in a village grew watermelons at the same time. Would the market collapse? I wrote to my colleague in the Islands, Dr Pita Tikae, a fruit and vegetable expert. Pita had been to Takwa this year and had also seen one of the villagers in Honiara looking for watermelon seed. The farmers of Takwa did so well with their watermelon last year that they are doing it again this year for the holiday season. There is now so much demand for their fine melons that the prices have stayed high. Now their only problem is getting enough seed. They are making their own money by adopting Osanti's innovation. Sometimes one person does make a difference.*

Related video

Elsewhere in the Asia-Pacific region, some people raise (other species of) crabs to make money.

Shushilan, BIID and mPower (2020) *Hardening Crabs in Floating Cages*. Hosted by Access Agriculture. 10 minutes.
www.accessagriculture.org/hardening-crabs-floating-cages

A version of the section above was previously published on www.agroinsight.com in 2014, by Jeff Bentley.

The Joy of Business

On 29 June in Cochabamba, I watched as 39 farmers' associations met with 183 businesses in a large rented ballroom, where tables just big enough for four people were covered in white tablecloths and arranged in a systematic grid pattern.

All day long the farmers and entrepreneurs huddled together in 25-minute meetings, scheduled one after the other, for as many as 15 meetings during the day, as the farmers explained the virtues of products like aged cheeses, shade-grown cacao and bottled mango sweetened with *yacón* (an Andean tuber, *Smallanthus sonchifolius*). Some businesses had come to buy these products, but others were there to sell the farmers two-wheeled tractors and other small machines.

Each association or business had filled out a sheet listing their interests and products. The organizer used computer software to match up groups by interest, and set a time for the meetings. The time was tracked by a large computerized clock projected on to the wall.

At the end of each of the 25-minute meetings, each table filled out a one-page form stating if they had agreed to meet for another business deal (yes, no, maybe), and, if so, when (within 3 months, or later), and the amount of the probable deal. By the end of the day, the farmers and the business people had agreed to do business worth 56 million Bolivianos (US$8.2 million).

Business representatives came from five foreign countries, Belgium, Peru, the Netherlands, Spain and Argentina, to buy groundnuts and other commodities. But most of the buyers and sellers were from Bolivia and only 6% of the trade was for export.

The meeting was self-financed. Each farmers' group paid US$45 to attend and each entrepreneur paid US$50. This is the ninth annual agro-business roundtable, so it looks like an institution that may last.

Business is a two-way street. For example, one innovative producer of fish sausages made deals to sell his fine products to hotels and supermarkets, but he also agreed to buy a machine to vacuum pack smoked fish, and made another deal to buy trout from a farmers' association.

With over 400 people lost in happy conversation on the ballroom floor, I barely noticed the three staff members on the side, sitting quietly at a table, typing up each sheet from each deal, using special software which allows the statistics to be compiled in real time. This will also help with follow-up. Two months after the roundtable, professionals from *Fundación Valles* (the event organizer) will ring up the group representatives with a friendly reminder: 'You are near the 3-month mark when you agreed to meet and buy or sell (a given product). How is that coming?'

Miguel Florido, who organized the event, explained that in previous years the roundtable brought in US$14 million in business, but that was mostly with banks and insurance companies, signing big credit deals or insurance policies. Now the amount of money has dropped a bit, but people are buying and selling tangible, local products, which is what the farmers want. It can be difficult and time-consuming for smallholders and entrepreneurs to meet each other, but with imaginative solutions buyers and sellers can connect.*

📹 Related video

Making mango crisps, or chips, is another promising business idea for small farms.

Biovision (2019) *Making Mango Crisps*. Hosted by Access Agriculture. 11 minutes.
www.accessagriculture.org/making-mango-crisps

---

*A version of the section above was previously published on www.agroinsight.com in 2017, by Jeff Bentley.

Commercializing Organic Inputs

As the world is waking up to address the challenges of environmental degradation and climate change, many countries realize that chemical fertilizers and pesticides are technologies of the past. While organic and ecological farmers use their ingenuity to keep pests and diseases at bay and to improve soil fertility with local inputs, the commercial sector has also seen the enormous potential to sell natural products.

Jeff, Ana, Marcella and I were welcomed by engineer Jimmy Ciancas, at Biotop, the commercial wing of the Foundation for the Promotion and Research of Andean Products (PROINPA – *Fundación para la Promoción e Investigación de Productos Andinos*), a Bolivian research agency headquartered in Cochabamba. We were impressed by the sophisticated technical set-up of the Biotop plant, where they mass produce a wide range of organic inputs such as probiotics. Biotop has invested years of research and development into its organic inputs, testing them all on farmers' fields before producing them commercially.

After analysing local soil samples in the laboratory, the most effective microorganisms are isolated and then mass multiplied. Besides beneficial lactic acid bacteria and yeasts, Biotop also produces bacteria and fungi that can kill insect pests or harmful fungi. In each of their four bioreactors, they can multiply 120 l of highly concentrated microorganisms once every 3 days. As it only takes 100 ml to spray a hectare, their current set-up provides organic inputs for 400,000 ha of agricultural land.

This technology has an enormous potential to boost organic and natural farming across the globe. It is also good to see a country like Bolivia making its own organic inputs, keeping some measure of independence from multinational corporations.

When we tell Jimmy Ciancas that we are making a farmer training video on biol, a fermented liquid fertilizer, we ask if PROINPA also makes this.

> It is one of the few products that we do not produce, because there is no profit to be made with biol. It requires too much work. Also, our commercial organic inputs need to be certified by SENASAG [the national agricultural health and food safety agency], so we need to have highly standardized products. If the label says that it contains one type of microorganism at a given concentration, the product needs to be as stated on the label.

I realize that these regulations are intended to standardize products, but they also limit the ability of a company to make more complex mixtures. A spoonful of soil has thousands of species of microorganisms (Baveye *et al.*, 2016), so limiting a preparation to just one species may do little to increase the diversity of a complex living soil community.

I wondered if biol could be useful to fight a soil pathogen, or to boost soil fertility, so I ask Jimmy Ciancas if biol is a useful technology for farmers.

> Yes, of course it is. It enriches the complex community of soil microorganisms with a variety of beneficial bacteria, fungi and yeasts.

As farmers mix legumes that are rich in plant hormones with the fresh manure of their animals, the microorganisms in biol made by one farmer will differ from a batch made by another.

I was glad to hear an expert confirm the usefulness of biol, an uncomplicated technology. The training video we are making on biol will be appreciated by farmers who want to make their own solutions.

The visit to the factory reminded us that bioinputs may be profitably made on different scales. A modern company with a state-of-the-art plant can refine specific, beneficial microorganisms and sell them in convenient bottles to thousands of farmers across the country. Meanwhile, farmers can make their own inputs with many microorganisms, which will also fight pests and improve the soil. Commercial and craft styles of making beneficial organisms will both be useful in the transition away from imported agrochemicals.*

Related video

Organic growth promoters are made from natural ingredients and can also contain herbal pest repellents.

MSSRF (2022) *Organic Growth Promoter for Crops*. Hosted by Access Agriculture. 16 minutes.
www.accessagriculture.org/organic-growth-promoter-crops

---

*A version of the section above was previously published on www.agroinsight.com in 2023, by Paul Van Mele.

References

Baveye, P.C., Berthelin, J. and Munch, J.-C. (2016) Too much or not enough: Reflection on two contrasting perspectives on soil biodiversity. *Soil Biology and Biochemistry* 103, 320–326.

Boonabaana, B., Musiimenta, P., Mangheni, M.N. and Ankunda, J.B. (2020) *Youth Realities, Aspirations, Transitions to Adulthood and Opportunity Structures in Uganda's Dryland Areas.* Report submitted to ICRISAT, Nairobi, Kenya.

Endris, G.S. and Hassan, J.Y. (2020) *Youth Realities, Aspirations, Transitions to Adulthood and Opportunity Structures in the Drylands of Ethiopia.* Report submitted to ICRISAT, Nairobi, Kenya.

Foale, S. (1999) Local ecological knowledge and biology of the land crab *Cardisoma hirtipes* (Decapoda: Gecarcinidae) at West Nggela, Solomon Islands. *Pacific Science* 53, 37–49.

GAOD (2023) Global Alliance for Organic Districts (GAOD) website. Available at: https://gaod.online/ (accessed 16 October 2024).

HLPE (2019) Agroecological and Other Innovative Approaches for Sustainable Agriculture and Food Systems that Enhance Food Security and Nutrition. Report by the High Level Panel of Experts on Food Security and Nutrition (HLPE), FAO, Rome, Italy. Available at: www.csm4cfs.org/summary-recommendations-hlpe-report-agroecology-innovations/ (accessed 16 October 2024).

Hurley, J.M. (2012) Recovery of the terrestrial crab *Cardisoma carnifex* after burrow disturbance. Available at: www.escholarship.org/uc/item/41z997bp#page-1 (accessed 16 October 2024).

Mácháč, J., Trantinová, M. and Zaňková, L. (2021) Externalities in agriculture: How to include their monetary value in decision-making? *International Journal of Environmental Science and Technology* 18, 3–20.

Mwaseba, D.L., Ahmad, A.K. and Mapund, K.M. (2020) *Youth Realities, Aspirations and Transitions to Adulthood in Dryland Agriculture in Tanzania.* Report submitted to ICRISAT, Nairobi, Kenya.

15 When Researchers and Farmers Collaborate

Abstract

Successful collaborations between researchers and farmers depend on mutual respect and shared goals. The case studies show that farmers and researchers in West Africa are collaborating to develop botanical methods to control the striga weed. Bangladeshi farmers grow pesticide-free vegetables for their own families, while relying on agrochemicals for commercial vegetables, a valuable lesson for pest control researchers. Bolivian farmers and researchers are collaborating to combat a new pest, the potato tuber moth, by combining traditional knowledge with scientific research. Andean farmers combine local weather forecasting methods with modern meteorological forecasts to make informed agricultural decisions. Innovation fairs bring together farmers, researchers and businesses to share ideas and develop practical solutions. Farmers in Bolivia are collaborating to develop effective organic alternatives to chemical fertilizers and pesticides. These examples show how to leverage the strengths of both farmers and scientists to find practical innovations.

A Common Ground

Farmers need new ideas, and researchers need data. When these two professional groups meet in the framework of collaborative or participatory research, it is often not clear who has to evolve in what direction: do farmers need to learn about research protocols, systematically collecting and analysing data, or do researchers need new ideas from farmers to guide their research agenda?

When grantees of the McKnight Foundation from West Africa met in 2020 in Montpellier, France, at a Community of Practice meeting to share experiences, it was refreshing to see how this network has over time taken ownership of some key values on doing research with farmers on agroecology, as

© Jeffery W. Bentley and Paul Van Mele 2025. *Agroecology in Practice: From Local Initiatives to Global Scaling Through Video* (J.W. Bentley and P. Van Mele) DOI: 10.1079/9781800628793.0015

a way to move towards a more just and equitable food system with care for the people and the planet.

Out of the more than 60 people from farmer organizations, NGOs, research institutes and universities from Mali, Burkina Faso and Niger, I was glad to run into some old friends. Ali Maman Aminou is a farmer and director of the federation of farmer unions in Maradi (FUMA Gaskiya), one of the main farmer organizations in Niger.

In 2011, Aminou was one of the 12 people we trained during a two-week intensive workshop on making quality farmer-to-farmer training videos. Ever since, Aminou has been using video in his interactions with the growing number of members, now some 18,000. The series of ten videos on integrated striga and soil fertility management that were developed with the International Crops Research Institute for the Semi-Arid Tropics (ICRISAT) and its partners were all translated into Hausa, which made it an ideal tool to trigger lively discussions with farming communities. Striga (*Striga* spp.) is a parasitic weed that attaches its roots to the roots of cereal crops, as such depriving the crop of the water and nutrients it needs (Mounde *et al.*, 2020). Aminou said,

> During one of the evenings that we showed the videos, one of the farmers spoke out and told us that he liked the videos, but that they had another technology to fight striga that was also efficient.

Aminou listened intently as the man went on to explain that farmers mix their millet seed with the powdery substance found around the seeds of the néré (*Parkia biglobosa*), a common tree across West Africa. When farmers sow millet, the néré powder apparently inhibits the striga seeds in the soil from germinating.

'This is amazing', I told Aminou. 'It would be great if you could turn this into a training video.' At that stage, it became apparent how much farmers and researchers had already begun to interact as equal players. Aminou swiftly turned to Salifou Nouhou Jangorzo, a lecturer from the University of Maradi in Niger, who had joined our discussion, and said, 'We need to find out more about this practice. We need all the details of how farmers do this.' Professor Salifou looked surprised at first; he had never heard of this practice before, but after 5 minutes of discussing with Aminou, he was convinced. It turns out that he was planning a survey on a labour-saving weeding technology and so he decided on the spot that he would add some questions about managing striga with néré to his survey.

Farmer-to-farmer training videos, like the ones in the striga series, inspire farmers to experiment with new ideas. They also give farmers confidence to openly share their real-life experiences, knowledge and practices. Through a functional network, these ideas can find their way back to researchers. In a progressive and collaborative research network, communication is not an end-product in itself, as Aminou has shown, but it feeds into a life of learning to make agriculture more resilient, profitable and responsive to farmers' needs.

Finding a common ground between researchers and farmers does not happen overnight; it needs a concerted and long-term effort.*

📽 Related video

This is one of a set of videos explaining the striga weed and how to control it, part of the series of videos on striga mentioned in this section.

Agro-Insight and ICRISAT (2016) *Striga Biology*. Hosted by Access Agriculture. 9 minutes. www.accessagriculture.org/striga-biology

---

*A version of the section above was previously published on www.agroinsight.com in 2020, by Paul Van Mele.

Tomatoes Good Enough to Eat

I was astounded years ago to learn that many farmers in Bangladesh had two completely different ways to grow vegetables. As my friend and colleague Harun-ar-Rashid told me, farmers sprayed pesticides as often as every other day on their commercial vegetables, yet grew a pesticide-free crop to eat with their families.

It's not that I doubted Harun's story. He's a careful observer and an experienced Bangladeshi agricultural scientist, but I wanted to find out more about this odd contradiction. How could farmers simply do without pesticides on crops that usually required a lot of spraying? Harun's explanation was that the farmers were worried about eating vegetables tainted with dangerous chemicals. But that assumed that there were viable alternatives to the intense use of pesticides.

In 2017, I got to see for myself how this double standard works. Paul and I were teaching a video scriptwriting course, and tagging along with some of my mature students. They were writing a video script on tomato late blight, the same vicious disease that also destroys potato crops. We were visiting family farmers who grew commercial vegetables in the village of Sordarpur in the south-west of Bangladesh, near Jessore. The farmers had received a lot of training from extensionists and had thoughtfully blended the new information with their own experience.

On their commercial fields, as soon as the farmers see late blight symptoms on tomatoes, they begin spraying with fungicides. The growers monitor the tomato crop constantly and spray often, especially when foggy days are followed by sun, which is perfect weather for late blight.

Farmers go to their commercial fields every day to check their tomatoes and prune diseased leaves with scissors. Then they clean the scissors with disinfectant to avoid spreading disease from plant to plant. Farmers can hire labour to do this in their commercial fields. They say that because of the fungicides, there are few diseased leaves in the commercial fields. The diseased leaves are collected in a bag or bucket to keep them from spreading disease to the healthy plants.

In their small home gardens, the farmers grow around ten plants and uproot the ones that get diseased instead of spraying them. The farmers said that about eight plants usually survive, enough to feed the family.

The farmers in Sordarpur graft their home garden tomatoes onto aubergine (eggplant) rootstocks. Partly this gives the tomatoes a stronger stem, but the farmers also think that grafting protects the tomatoes from disease, although they are not sure why. (Grafting can provide disease-resistant rootstocks to manage a disease like late blight, which is transmitted in the soil and through the air.)

Insect pests can also be a problem. In the home gardens, farmers control insect pests (such as aphids and fruit flies) by hanging up plastic pots painted yellow and coated with engine oil. The fruit flies are attracted to the colour yellow and get stuck in the oil. The farmers are also starting to use sex pheromone traps, trying out this new practice mostly in the home gardens.

They make organic pesticides with mustard seed oil, which is used only or mainly in the home gardens. Store-bought chemical insecticides are used in the commercial fields.

For commercial produce, farmers spray pesticides in response to market demands for healthy-looking fruits and vegetables. But if the farmers think that the resulting produce is too dangerous to eat, perhaps consumers should take note and demand healthier vegetables. No doubt when the market demands pesticide-free produce, researchers will be even more motivated to come up with effective alternatives, such as sex pheromone traps, and commercial growers will adopt them.

Agricultural scientists need to pay attention to farmers' alternative pest control techniques. For example, grafting could be done by the enterprising farmers who prepare seedlings as a business. Start with what farmers successfully do to grow their own organic vegetables and explore further. It is ironic that city people are eating more fruits and vegetables for their health, without knowing if those crops received toxic doses of chemicals. Many farmers are unwilling to eat the conventional produce that they produce; this says that something is wrong with conventional pest control. Pesticide-free vegetables may be lower yielding and take more work to produce, but could be profitable if consumers are willing to pay more for them, and if research and extension would support more alternatives to chemicals.*

📹 **Related video**

Farmers in India explain some alternative disease-control methods.

Pagar, A. and WOTR (2023) *Controlling Wilt Disease in Pigeon Pea.* Hosted by Access Agriculture. 15 minutes.
www.accessagriculture.org/controlling-wilt-disease-pigeon-pea

A version of the section above was previously published on www.agroinsight.com in 2017, by Jeff Bentley.

Zoom to Titicaca

COVID-19 may be the world's most spectacular emerging disease, but agriculture has its own new pests and diseases. Fortunately, collaboration between agronomists and farmers can offer solutions, as I saw in a recent meeting on the shores of Lake Titicaca.

This was in 2021 and Bolivia was still in COVID-19 lockdown, so we met on Zoom, but I was struck by how much this virtual meeting resembled face-to-face meetings I have attended with farmers and agronomists.

Ing. Sonia Laura, a researcher from Promotion of Sustainability and of Shared Knowledge (PROSUCO – *Promoción de la Sustentabilidad y Conocimientos Compartidos*), a Bolivian NGO, who works closely with farmers, had driven out to the village of Iquichachi, a couple of hours from her base in La Paz. Sonia set up the call on her laptop, and the farmers (Sra. Cristina, Sra. Arminda, Sr. Juan, Sr. Paulino, Sr. Zenobio and Sr. Fidel) all managed to squeeze onto the screen. Bundled up in coats and hats against the high Andean cold, they explained how several years ago, they noticed a new worm eating the potatoes they store at home.

The moth lays its eggs on stored potatoes and on potato plants in the field. The eggs hatch into caterpillars that go back and forth: from field to home in the harvest, and from storage to field with the seed.

The farmers showed some graphs of data they had been collecting with Sonia, under advice from Ing. Reinaldo Quispe, an agronomist from the Foundation for the Promotion and Research of Andean Products (PROINPA – *Fundación para la Promoción e Investigación de Productos Andinos*), a Bolivian research agency, who joined the call from his office in La Paz. Reinaldo and the farmers had been using the sex scent (pheromone) of female moths to attract and trap the male moths (Figs 15.1 and 15.2).

Each moth species has its own unique sex pheromone. Reinaldo had identified the pests, two species of tuber moths (*Phthorimaea operculella* and *Symmetrischema tangolias*) that belong to the Gelechiidae, a family that specializes in eating stored foods. Both species are native to the Andes, but usually found in the lower, warmer valleys (Dangles *et al.*, 2010; Crespo-Pérez *et al.*, 2015).

Raúl Ccanto, an agronomist from the NGO Yanapai from Peru, joined us. Raúl explained that Peruvian farmers had suffered from these two moths for many years. Over the years of working with the farmers, Yanapai and others had developed some practical solutions.

As Raúl explained, select the seed carefully. When you take seed from the house to plant in the field, make sure that you only plant healthy tubers, not the ones full of worms.

Also rotate your crops. 'This is something you farmers have always done, but it's important to say that it is a good thing.' Growing potatoes one

Fig. 15.1. Checking for fallen tuber moths in a pheromone trap.

year, followed by other roots and tubers – such as oca (*Oxalis tuberosa*) and *papalisa* (*Ullucus tuberosus*), which are not of the potato family – and then other legumes and cereals, helps to keep the soil free of potato pests.

Raúl's PowerPoint included the results of experiments done in collaboration with Peruvian farmers where they tried various ways to manage the moths in stored seed potato. One idea that worked well, and was also cheap, was to dust healthy seed potatoes with talc, which keeps the moths from laying their eggs in potatoes. The talc worked almost as well as malathion, the insecticide.

Raúl skipped lightly over the malathion, barely mentioning it, and for good reason. He had included the chemical treatment in the experiment as a comparison, but he was not promoting it. As Reinaldo explained, farmers often prefer insecticides and use them even in stored potatoes, which one should not do.

In fact, medical schools in Bolivia teach their third-year students to diagnose and treat malathion poisoning, because it is so common. 'This is something you'll see', the older doctors tell their students (personal communication, Clara Bentley, Bolivia).

With any new pest or disease, it's important to know where it came from. Raúl explained that the moths may have recently colonized the cold Altiplano, not just because of climate change, but also because people are bringing wormy seed in from fairs in distant parts of the country. And they are growing more potatoes. As more of the land is planted more often and

Fig. 15.2. The larvae of the potato tuber moth can also bore into the stem of potato plants, causing them to wilt.

over larger areas, to meet market demand, a more attractive environment is created for potato pests.

Yes, the farmers agreed, potatoes are being grown more often. And that is why it is crucial for scientists and farmers to put their heads together, to confirm useful ideas from different perspectives.

The farmers wanted to know if there was something they could apply to their potatoes to kill the moth. Raúl and Reinaldo both explained that there is no one thing that will manage the pest. It will have to be managed by rotating crops and by selecting healthy seed. Other ideas like dusting the potatoes with talc will also help. The good news is that the moths can be managed.

It may be in human nature to yearn for simple solutions. Many of us have simply wished that COVID-19 would go away, and that things would go back to normal. Like COVID-19, managing the tuber moth will require several good ideas, well explained, widely shared and applied.

In this case, the new information motivated the farmers to set up their own experiments. Sonia told me that after our call, the farmers met to reflect and take action. They decided that each one of them would select their seed, clean their potato storeroom and sprinkle talc on the selected seed. They will keep using the pheromone traps, among other things. Later, they will explain these practices to their other community members, to take action as a group.

Two years later, we produced the video below with trained farmers, in collaboration with PROINPA.*

📹 **Related video**

This video is on the tuber moth, with versions in Spanish and two native languages of Bolivia (Quechua and Aymara), as well as French and English.

Agro-Insight and PROINPA (2023) *Managing the Potato Tuber Moth*. Hosted by Access Agriculture. 15 minutes.
www.accessagriculture.org/managing-potato-tuber-moth

---

*A version of the section above was previously published on www.agroinsight.com in 2021, by Jeff Bentley.

Predicting the Weather

Most city dwellers are only interested in short-term weather forecasting. Will it rain over the weekend when we plan to invite friends to a barbecue? Do I need to carry an umbrella or wear a coat tomorrow? Fortunately for urbanites, TV, radio and web-based services provide short-term forecasts.

Farmers are interested in short-term weather forecasting too, but also in long-term predictions. Knowing which week the rains will start is crucial for deciding when to plant rain-fed crops. Knowing how much it will rain helps farmers choose whether to plant on high or low ground.

I learned this recently from Edwin Yucra, a researcher at *Universidad Mayor de San Andrés* (UMSA), the public university of San Andrés in La Paz, Bolivia. Edwin has spent years working with Andean farmers on the Bolivian Altiplano, helping them to make use of weather forecasts based on the latest scientific data. For example, not long ago, Edwin noticed that there was an unexpected rain forecast for 2 or 3 days' time. Farmers usually like rain, but not on this occasion. The farmers he works with were about to freeze-dry potatoes into *chuño*, when dry nights are essential. To warn the farmers, Edwin didn't have to meet with them. He let them know on social media. The farmers were able to delay making *chuño* and save their potatoes from rotting.

Scientific weather forecasting is not particularly accurate over a whole year. This leaves farmers more or less to their own devices. One group of master Andean farmers, called the *Yapuchiris* (which means 'farmers' in Aymara), is paying attention to long-term weather forecasting (Quispe *et al.*, 2018). During the dry season, the *Yapuchiris* notice the behaviour of animals, plants or stars. For example, birds nesting on high ground are interpreted as a sign of a wet year, while low-lying nests suggest a coming drought (Choquetopa Rodríguez, 2021).

The *Yapuchiris* write down their meteorological predictions and then painstakingly record the weather every day for the next year, to see if their forecasts are accurate. The *Yapuchiris* use a paper form which they and their partners at PROSUCO (an NGO) have been perfecting since the early 2000s. They use a large chart called a *Pachagrama*. They coined this term by blending the Aymara word for 'earth' and 'weather' (*pacha*) with the Spanish ending *-grama* (as in *telegrama*). The 'Earth-gram' includes 365 columns for each day of the year and rows for different kinds of weather (sun, wind, rain, hail, etc.). The *Yapuchiris* draw a dot in each row every day to add further information. For example, a dot placed higher in the sun column means a sunny day and a lower dot is a cloudy day. Later the dots can be connected to draw a graph of the year's weather.

Modern meteorology tracks weather in the short term, but is less accurate after a week. A ten-day weather forecast is only right about half the time (NOAA, 2023). However, local people make weather predictions over several months (Orlove *et al.*, 2000). These two efforts are different, but farmers value both of them and will use them to see what the weather will be like this week, and this year. Collaboration with Bolivian researchers is helping the *Yapuchiris* to get some recognition and respect for their knowledge. Some research has suggested that local weather forecasts are largely valid (Claverías, 2020). More needs to be done to validate these predictions, comparing farmer forecasts with actual weather records. That would be of benefit to farmers, and to meteorologists.*

▶ Related video

This video from Bolivia explains how to blend ethno-meteorology and modern weather forecasting.

Agro-Insight, UMSA and PROSUCO (2019) *Forecasting the Weather*. Hosted by Access Agriculture. 15 minutes.
www.accessagriculture.org/forecasting-weather

---

*A version of the section above was previously published on www.agroinsight.com in 2018, by Jeff Bentley.

Five Heads Think Better Than One

Innovation fairs are becoming a popular way to showcase agricultural invention and to link some original thinkers with a wider community.

On 28 June 2017, I was at an innovation fair in Cochabamba. It was held in a ballroom that is usually rented for weddings and big parties, but with some tweaking it was a fine space for farmers and researchers to meet. Each organization had a table where they could set out products or samples, with their posters displayed behind the presenters.

For example, at one table, I met a dignified white-haired agronomist, Gonzalo Zalles, who explained his work with 'deep beds' for raising healthy, odourless pigs. I told Mr Zalles about some pigs I had seen in Uganda, but he explained that he makes a slightly more sophisticated bed. He starts by digging a pit, then adding a thin layer of lime to the base, followed by a layer of sand. In Uganda, some innovative farmers raise pigs on wood shavings, but Zalles uses rice husks as the final layer. He says they are more absorbent than wood shavings.

I asked if he added Effective Microorganisms® (a trademarked brand of yeast and other microbes that are used widely, in odourless pigpens and to make bokashi). But no, in Bolivia, swine farmers are using a mix of bacteria and yeast called BioBull, which is made by Biotop, a subsidiary of the PROINPA Foundation in Cochabamba.

At a nearby stall, I caught up with José Olivera of Biotop, who was displaying not just BioBull, but other biological products as well, including insecticides and fungicides for organic agriculture. José travels all over the Bolivian Altiplano selling these novel inputs to farmers. The Panaseri Company, in Cochabamba, collaborates with PROINPA to produce food products from the lupin bean, packaged for supermarkets under the brand Tarwix (from *tarwi*, the local name for lupin). At the Panaseri stand, Norka Ojeda, a PROINPA communicator, explained that the Tarwix factory buys lupin beans (*Lupinus mutabilis*) from farmers and washes out the poisonous alkaloids, rendering the nutritious *tarwi* safe to eat. By the 2020s, Tarwix was being sold in Bolivia.

Linking researchers to farmers' associations and companies seems to be bearing fruit. Raising swine without the bad smell is crucial for keeping livestock near cities, where it is easy to get supplies and the market for the final product is nearby. Packaging traditional foods like lupin beans also opens new markets for crops produced by smallholders. As of 2024, those packaged lupins were still being sold in Bolivia. Many heads think better than one.*

■ Related video

Involving women in agricultural extension can also be a way to connect farmers and researchers.

Agro-Insight (2016) *Women in Extension*. Hosted by Access Agriculture. 16 minutes. www.accessagriculture.org/women-extension

---

A version of the section above was previously published on www.agroinsight.com in 2017, by Jeff Bentley.

Organic Leaf Fertilizer

To encourage organic farming, PROSUCO, a Bolivian NGO, teaches farmers how to make alternatives to chemical pesticides, including: i) sulfur-lime: water boiled with sulfur and lime and used as a fungicide (some farmers also find that it is useful as an insecticide); ii) Biofoliar, a fermented solution of cow and guinea pig manure, chopped lucerne (alfalfa), ground egg shells, ash, and some shop-bought ingredients: brown sugar, yoghurt and dry active yeast (Quispe and Laura, 2020). After a few months of fermenting in a barrel, the biofoliar is strained and can be mixed in water to spray onto the leaves of plants.

Conventional farmers often buy foliar (leaf) chemical fertilizer, designed to spray on a growing crop. But this foliar chemical fertilizer is another source of impurities in our food, because the chemical is sprayed on the leaves of growing plants such as lettuce and broccoli.

I was with PROSUCO in 2023, making a video in Cebollullo, a community in a narrow, warm valley near La Paz, with Paul and Marcella from Agro-Insight. These organic farmers usually mix biofoliar together with sulfur-lime. They rave about the results. The plants grow so fast and healthy, and these homemade remedies are much cheaper than the chemicals from the shop.

The mixture does seem to work. One farmer, doña Ninfa, showed us her broccoli. There were cabbage moths (*Plutella xylostella*) flying around it and landing on the leaves. These little moths are the greatest cabbage pest worldwide, and also a broccoli pest. Doña Ninfa had sprayed her broccoli with sulfur-lime and biofoliar. I saw a little *Plutella* damage on the leaves, but I couldn't find any of their larvae, little green worms. So whatever doña Ninfa was doing, it was working.

I do have a couple of questions. Biofoliar provides nutrients when used as a foliar fertilizer, but I wonder if it also has beneficial microorganisms that help the plants? To know that, we would have to assay the microorganisms in the biofoliar, before and after fermenting it. Then we would need to know which microbes are still alive after being mixed with sulfur-lime, which is designed to be a fungicide, i.e. to kill disease-causing fungi. The mixture may reduce the number of microorganisms, but this would not affect the quantity of nutrients for the plants. Farmer-researchers of Cebollullo have been testing different ratios of biofoliar and sulfur-lime to develop the most efficient control while reducing the number of sprays, to save time and labour. This is important to them because their fields are often far from the road and far from water.

This is not a criticism of PROSUCO, but there needs to be more formal research, for example, from universities, on safe, inexpensive, natural fungicides and fertilizers that farmers can make at home, and on the combinations of these inputs. A recent review found that published research on liquid ferments and biol is uneven. However, at least some liquid organic fertilizers do have beneficial microorganisms and plant hormones, but more research is needed on these products (O'Neill and Ramos-Abensur, 2022).

Agrochemical companies have all the advantages. They co-opt university research. They have their own research scientists as well. They have advertisers and a host of shopkeepers, motivated by the promise of earning money. Organic agriculture has the good will of the NGOs, working with local people, and the creativity of the farmers themselves. More support for research on organic farming and agroecology would make a difference.*

▶ **Related video**

This is the video on making enriched biofertilizer referred to in this section.

Agro-Insight and PROSUCO (2023) *Making Enriched Biofertilizer*. Hosted by Access Agriculture. 16 minutes.
www.accessagriculture.org/making-enriched-biofertilizer

---

*A version of the section above was previously published on www.agroinsight.com in 2023, by Jeff Bentley.

References

Choquetopa Rodríguez, B. (2021) *Indicadores Naturales para Pronosticar el Tiempo en el Sur de Oruro, Bolivia (Natural Indicators to Forecast the Weather in the South of Oruro, Bolivia)*. Bentley, J.W. (ed.). Agro-Insight and the McKnight Foundation, Cochabamba, Bolivia.

Claverías, R. (2020) Conocimientos de los campesinos andinos sobre los predictores climáticos: elementos para suverificación. Trabajo expuesto en el Seminario-Taller NOAA, Missouri [Andean peasant knowledge of weather predictors: elements for their verification. Paper read at the NOAA Seminar-Workshop, Missouri].

Crespo-Pérez, V., Régnière, J., Chuine, I., Rebaudo, F. and Dangles, O. (2015) Changes in the distribution of multispecies pest assemblages affect levels of crop damage in warming tropical Andes. *Global Change Biology* 21, 82–96.

Dangles, O., Carpio, F.C., Villares, M., Yumisaca, F., Liger, B. *et al.* (2010) Community-based participatory research helps farmers and scientists to manage invasive pests in the Ecuadorian Andes. *AMBIO* 39, 325–335.

Mounde, L.G., Anteyi, W.O. and Rasche, F. (2020) Tripartite interaction between *Striga* spp., cereals, and plant root-associated microorganisms: A review. *CABI Reviews* 15, 005.

NOAA (2023) How Reliable Are Weather Forecasts? The National Oceanic and Atmospheric Administration (NOAA), United States Department of Commerce. Available at: https://scijinks.gov/forecast-reliability/ (accessed 16 October 2024).

O'Neill, B. and Ramos-Abensur, V. (2022) A Review of the State of Knowledge and Use of Liquid Ferments and Biol in the Andes. Report submitted by Rikolto and the University of Michigan to the McKnight Foundation. Available at: www.ccrp.org/wp-content/uploads/2022/10/A-review-of-the-state-of-knowledge-and-the-use-of-liquid-ferments-and-biol-in-the-Andes.pdf (accessed 16 October 2024).

Orlove, B.S., Chiang, J.C.H. and Cane, M.A. (2000) Forecasting Andean rainfall and crop yield from the influence of El Niño on Pleiades visibility. *Nature* 403, 68–71.

Quispe, M. and Laura, S. (2020) *Bioinsumos: Un Aporte a la Resiliencia de los Sistemas Productivos (Bioinputs: A Contribution to the Resilience of Farming Systems)*. PROSUCO, La Paz, Bolivia.

Quispe, M., Laura, S. and Baldiviezo, E. (2018) *Yapuchiris: Un Legado para Afrontar los Impactos del Cambio Climático (Yapuchiris: A Legacy to Face the Impacts of Climate Change)*. PROSUCO, Cooperación Suiza, La Paz, Bolivia.

PART 4

Communicating with Farmers through Video

Videos can be made to teach agroecology cross-culturally, as discussed in Parts 1–3. Agro-Insight is a small company that produces videos with empowered farmers, for other farmers. Access Agriculture, a non-profit, supports South–South exchange and last-mile delivery of farmer-to-farmer training videos. Access Agriculture hosts videos made by Agro-Insight, and others. Part 4 describes some of our experiences of creating videos with farmers and examples of how farmers have innovated after watching the videos. Bureaucrats sometimes complain to us that farmer learning videos need to be 'filmed over again in our country'. Civil servants seem to think that farmers in South America won't understand a video filmed in South Asia. But they do. Farmers and organizations enjoy and benefit from videos filmed in other regions (for example, Bentley and Van Mele, 2011; Bentley *et al.*, 2013, 2019, 2022; Van Mele *et al.*, 2013, 2016; Karubanga *et al.*, 2017; Ongachi *et al.*, 2017; Maredia *et al.*, 2018; Zoundji *et al.*, 2018, 2020; Bede *et al.*, 2020; Gouroubera *et al.*, 2023).

References

Bede, L., Okry, F. and Vodouhe, S.D. (2020) Video-mediated rural learning: Effects of images and languages on farmers' learning in Benin Republic. *Development in Practice* 31, 59–68.

Bentley, J.W. and Van Mele, P. (2011) Sharing ideas between cultures with videos. *International Journal of Agricultural Sustainability* 9, 258–263.

Bentley, J.W., Van Mele, P. and Musimami, G. (2013) The Mud on Their Legs – Farmer to Farmer Videos in Uganda. MEAS Case Study # 3. Available at: https://agroinsight.com/downloads/Articles-Agricultural-Extension/2013_AE5_MEAS-CS-Uganda-Farmer-to-Farmer-Videos-Uganda-BentleyJ- and%20PVanMele-July%202013.pdf (accessed 16 October 2024).

Bentley, J.W., Van Mele, P., Barres, N.F., Okry, F. and Wanvoeke, J. (2019) Smallholders download and share videos from the internet to learn about sustainable agriculture. *International Journal of Agricultural Sustainability* 17, 92–107.

Bentley, J.W., Van Mele, P., Chadare, F. and Chander, M. (2022) Videos on agroecology for a global audience of farmers: An online survey of Access Agriculture. *International Journal of Agricultural Sustainability* 20, 1100–1116.

Gouroubera, M.W., Moumouni, I.M., Okry, F. and Idrissou, L. (2023) A holistic approach to understanding ICT implementation challenges in rural advisory services: Lessons from using farmer learning videos. *Journal of Agricultural Education and Extension* 30, 213–232.

Karubanga, G., Kibwika, P., Okry, F. and Sseguya, H. (2017) How farmer videos trigger social learning to enhance innovation among smallholder rice farmers in Uganda. *Cogent Food & Agriculture* 3, 1368105.

Maredia, M.K., Reyes, B., Ba, M.N., Dabire, C.L., Pittendrigh, B. *et al.* (2018) Can mobile phone-based animated videos induce learning and technology adoption among low-literate farmers? A field experiment in Burkina Faso. *Information Technology for Development* 24, 429–460.

Ongachi, W., Onwonga, R., Nyanganga, H. and Okry, F. (2017) Comparative analysis of video mediated learning and farmer field school approach on adoption of striga control technologies in Western Kenya. *International Journal of Agricultural Extension* 5(1), 1–10.

Van Mele, P., Bentley, J.W., Harun-ar-Rashid, M., Okry, F. and van Mourik, T. (2016) Letting information flow: Distributing farmer training videos through existing networks. *Indian Journal of Ecology* 43, 545–551.

Van Mele, P., Wanvoeke, J., Rodgers, J. and McKay, B. (2013) Innovative and effective ways to enhance rural learning in Africa. In: Wopereis, M., Johnson, D., Ahmadi, N., Tollens, E. and Jalloh, A. (eds) *Realizing Africa's Rice Promise*. CABI Publishing, Wallingford, UK, pp. 366–377.

Zoundji, G.C., Okry, F., Vodouhê, S.D. and Bentley, J.W. (2018) Towards sustainable vegetable growing with farmer learning videos in Benin. *International Journal of Agricultural Sustainability* 16, 54–63.

Zoundji, G.C., Okry, F., Vodouhê, S.D., Bentley, J.W. and Witteveen, L. (2020) Commercial channels vs free distribution and screening of learning videos: A case study from Benin and Mali. *Experimental Agriculture* 56, 544–560.

16 Making Learning Videos with Farmers

Abstract

Making videos with farmers highlights their agroecological innovations. This helps to counter the nagging persistence of top-down extension models, which often prioritize government and corporate agendas. Videos filmed with farmers on the Bolivian Altiplano highlighted their centuries-old knowledge of reading the signs of nature to predict the weather. Validating a factsheet on intercropping pigeon peas with maize in Malawi allowed the local researchers to incorporate farmers' knowledge into their work. Combining scientific research on aflatoxins with local women's experiences in Mali was crucial for making effective training videos. On a video from the Peruvian Andes, farmers explained how they ruined some of their land with chemical fertilizers, prompting the community to innovate with soil restoration techniques. Filming videos with farmers helps to show respect for their knowledge and experiments, and to share their ideas with farmers across the world.

Top-Down Extension on the Rise?

Despite more than three decades of investments in participatory approaches, top-down extension with blueprint recommendations is gaining ground again. Why is it so hard to stamp out such denigrating, disempowering practices that see farmers as passive takers of advice and obedient producers of food?

While working in Vietnam in 1997, roughly a decade after the government established a more liberal market economy with its *Đổi Mới* reform policy, my Canadian friend Vincent often shared his frustrations. As he deployed the tools of participatory rural appraisal (PRA) to assess the priority development needs of rural communities, vegetables often emerged as number one. But as he concluded the full day's exercise by asking the villagers what they wanted to work on, they always said 'rice'. It drove Vincent nuts, as there was no way he could justify that to his NGO back home. As

DOI: 10.1079/9781800628793.0016

rice was still set as a priority by the local authorities, people had put their personal aspirations aside and abided by government policies.

All states throughout history have relied on making people follow rules and pay taxes. In James Scott's book *Against the Grain*, he writes about the early development of agriculture, starting some 10,000 years ago (Scott, 2017). During the first several millennia of plant and animal domestication, early farmers and pastoralists continued to hunt and gather wild plants, leaving them with plenty of leisure time and an incredibly diverse and healthy diet as they practised sustainable agriculture for 4000 or 5000 years (Scott, 2017).

When the first states emerged some 6000 years ago, all this began to change. State elites collected tax as a share of the harvest or as forced labour (or both). As wheat, maize and rice need to be harvested at one particular time and can be stored, the early states forced farmers to grow more of these cereal crops. The first writings were not poems or epic stories, but accounts with names of people and taxes paid or other transactions. Rigid instructions on how to manage the crops allowed the tax collector to estimate yields and to calculate how much tax they could collect. Top-down extension is as old as the very first states. Crop diversity declined as people worked harder and ate less (Scott, 2017).

So, despite the more recent, huge public investments and overwhelming evidence of the benefits of participatory approaches, whether farmer field schools (FFS), community seed banks or participatory technology generation, development practitioners are up against a difficult enemy: a pushy state influenced by large corporations and tech companies that want to tell farmers what to do.

Over the past decade, non-traditional extension service providers like telecommunications companies, such as Safaricom, software developers and digital service providers, such as the internet giant Alibaba in China (Stone, 2022), have entered the scene. These developments are backed by donor agencies and philanthropists, such as the Bill and Melinda Gates Foundation and the Syngenta Foundation, who support digital extension as a way to shape the future of farming (Mann and Iazzolino, 2021).

Digital service providers may be able to provide pretty accurate information on market prices (Liao, 2018) and weather information or early warning systems, but none have combined this with agroecological advice (Burns *et al.*, 2022).

In this golden age of tweets, farmer advice is often summarized in short, simple text messages and, by doing so, digital service providers play back into the hands of those governments and companies who believe they have a right to control rural folks, under the guise of providing maximum benefits to farmers and reducing poverty. In 2019, Microsoft and the Alliance for Green Revolution in Africa (AGRA) co-created the AgriBot as a digital solution for localized extension and advisory services for smallholder farmers. The AgriBot works through inclusive omni-channel application experiences like Short Message Service (SMS) and WhatsApp for prioritized value chains, in partnership with companies like Bayer, Yara, Syngenta and Nestlé (Microsoft News Center, 2022).

Short, blunt messages are better for promoting agrochemicals than for sharing complex agroecological principles and supporting food sovereignty. While simple SMS messages can be offered in local languages, video will become an increasingly important format to engage farmers in active learning, with images and verbal discussion from fellow farmers. In video, the audience can read the images and listen to explanations by fellow farmers, plus viewers can go back and watch the video again and discuss it with their friends and family (Bentley *et al.*, 2016; Van Mele *et al.*, 2016). This gives video a depth and a subtlety that can't be tweeted.

But video can also be deployed to trigger farmers to adopt commercial seed, fertilizers and pesticides, and lure smallholder farmers into a system of surveillance agriculture. Digital Green, which is commonly portrayed as an organization promoting participatory video, was launched by Microsoft Research in 2006 and was later spun off as an independent non-profit, mainly funded by the Bill and Melinda Gates Foundation. With Microsoft's support, Digital Green has developed Farmstack – a platform for sharing farmer data and monitoring farmer compliance (Stone, 2022). Recent trends in artificial intelligence, and how Digital Green are deploying it, have raised strong concerns among civil society in Africa (Pascal *et al.*, 2023).

It is a rare digital service that understands farmers and responds to their needs in a non-directive way, that respects indigenous innovations, practices and knowledge and that encourages farmer experimentation. In a global review on digital tools to support agroecological transitions, more than 244 candidate tools were identified, of which 61 were then selected for a full review based on web searches, expert interviews and platforms. While all tools addressed some functions related to climate change mitigation or adaptation, only one tool (Access Agriculture) captured a significant number of agroecology functions (Burns *et al.*, 2022).

Anthropologist Paul Richards described small-scale farming as a type of performance whereby farmers learn by experimentation and adapt their behaviour to reach certain goals (Richards, 1989). To support diverse and healthy food systems, digital extension approaches will need to encourage experimentation and farmer-to-farmer learning across borders.

Modern states that see farmers as citizens, not as subjects, will need to explore many forms of participatory extension, and not simply try to digitize top-down approaches, which will never appeal to farmers. However, incentives to lure farmers into a model of industrial agriculture come in multiple forms. Given the growing influence of large corporations on governments, the role of civil society networks like the Alliance for Food Sovereignty in Africa (AFSA) that promote agroecology and influence policy makers cannot be underestimated. Farmers need to be protected by appropriate policies concerning digital services, as some may lead to distortions of competition and deteriorations of farmers' autonomy (Verdonk, 2019).*

> ### 📹 Related video
> ---
> The main limitation to using videos in cross-cultural learning is that farmers have to be able to understand the video. Hence, the ability to develop subsequent local language versions is crucial. The following video was filmed in Spanish with a women's group in Ecuador, but a year later is already available in seven other languages.
>
> Agro-Insight, CIP and INIAF (2022) *Managing Seed Potato*. Hosted by Access Agriculture. 16 minutes.
> www.accessagriculture.org/managing-seed-potato

*A version of the section above was previously published on www.agroinsight.com in 2021, by Paul Van Mele.

Let Nature Guide You

Farmers need to take decisions every day. Smallholders living in remote areas often have no one to turn to, to ask for advice. Nobody tells them which new crops to try, or when the soil will be moist enough to plant. In harsh environments, predicting the weather correctly can make the difference between harvesting a crop or harvesting nothing at all.

As always, when producing farmer training videos, we are fortunate to interact with farmers who are willing to share their knowledge and experiences. The NGO Promotion of Sustainability and of Shared Knowledge (PROSUCO – *Promoción de la Sustentabilidad y Conocimientos Compartidos*) has organized a network of 70 *Yapuchiris* (expert farmers). They receive some training, but mostly they are encouraged to share their own knowledge with other farmers (Quispe *et al.*, 2018). In the southern Altiplano of Bolivia, one of the *Yapuchiris*, Bernabé Choquetopa, explains that if frost hits your quinoa, you can lose your crop in the space of a day, all past efforts being in vain. He guides us in the brushland and shows us a local bush (*Fabiana densa*) called *tara t'ula* in the Aymara language. 'This plant doesn't like the cold very much, so if you find many of these plants, it is a good place to build your farm house, your corral to keep your llamas and grow your crop.'

But even if your farm is well located, frost can strike. So, don Bernabé has many other natural indicators to inform him about what actions to take. 'If the lizard makes a fresh house, it will rain tomorrow, but if it starts to close its burrow, it will freeze that night. I then collect *t'ula* plants and burn them in my quinoa field from 3 to 5 AM so that the frost will not settle on my crop', he continues.

Apart from observing plants and animals, don Bernabé also reads the clouds and wind. Amazingly, winds in June and July already tell him how the next rainy season that starts in January will be. Having arrived at a large sand dune, he points to the pattern of vertical ridges blown into the side of the dune. 'If the lines are some 10 cm apart, the rains will come close to each other and we will have a good harvest. But if they are further apart, the rains will also be sparser and our crop will suffer.'

Don Bernabé has written a book about these natural weather indicators (Choquetopa Rodríguez, 2021). As he shows us around the landscape, he proudly carries his book with colour photographs that clearly explain all the natural indicators he knows. Reading nature is a skill that requires spending a lot of time outdoors, observing natural phenomena.

The next few days we met some other extraordinary *Yapuchiris*, each sharing their knowledge with us in front of the camera. It is exciting to be part of this and at the same time an eye-opener as to how much industrial agriculture in the West has become disconnected from nature.

With climate change, the need to build on local knowledge will grow in importance.

In the words of don Bernabé:

Well, these plants and animals are more intelligent than the human being. They know how to live in this land and they know it perfectly. For that reason, it is necessary not to lose this knowledge and that the young people should keep practising this ancestral knowledge that is so rich.*

▶ Related video

Don Bernabé shares his knowledge and experiences in this video.

Agro-Insight, UMSA and PROSUCO (2019) *Forecasting the Weather*. Hosted by Access Agriculture. 15 minutes.
www.accessagriculture.org/forecasting-weather

A version of the section above was previously published on www.agroinsight.com in 2019, by Paul Van Mele.

The Rules and the Players

When we train local organizations, writing factsheets and validating them with farmers ensures the main ideas for the video are right, as illustrated by this account from a workshop Jeff and I held in Malawi in 2014.

Four groups of trainees went out to validate their factsheets with farmers (Fig. 16.1). No matter who you write for, one has to understand the audience and learn to listen. Most of our trainees were young people working with radio or TV, and had little knowledge on farming. While Jeff joined the group that wrote a factsheet on feeding dairy cows, I joined the pigeon pea group. We drove to Kaisi village, about 20 minutes from Kasungu, where the workshop took place. Farmers in northern Malawi grow maize as a staple crop and tobacco for cash. A few days earlier, coming from Lilongwe, it struck us

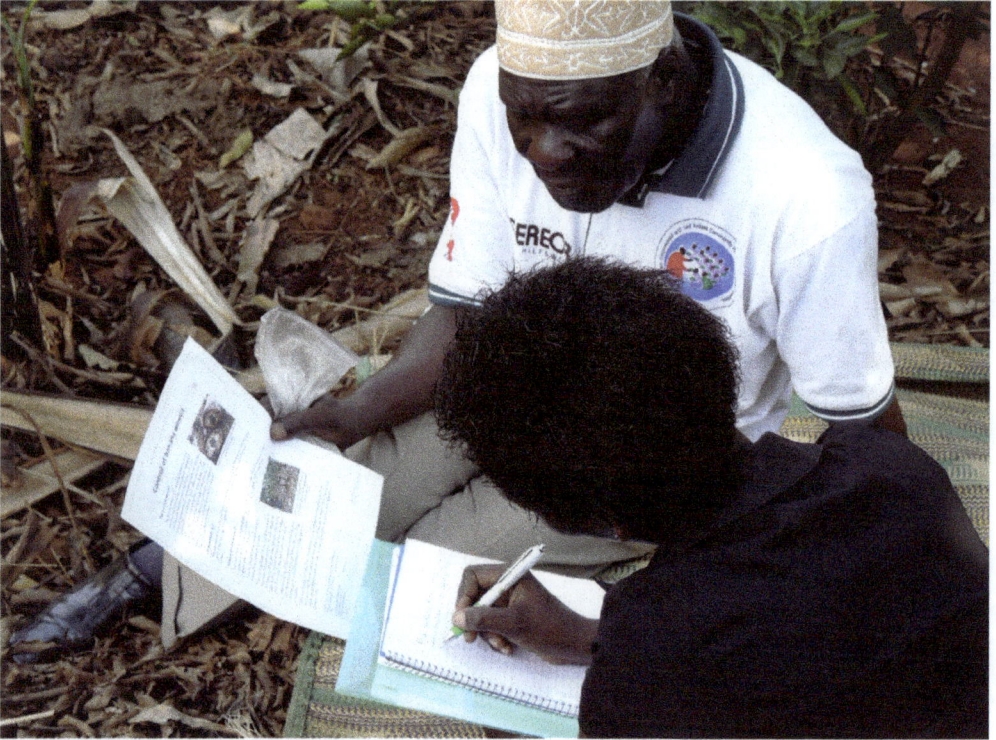

Fig. 16.1. Discussing a draft factsheet with farmers to avoid surprises when developing a learning video.

how all the farms were growing so much maize. Farmers only recently began growing pigeon pea, so I wondered how much experience farmers need to have with a crop before they can form useful insights. It turns out that farmers learn a lot, fast, but they still have major gaps in their knowledge.

Agriculturist Grace Tione had written her factsheet in English, but she read it out loud in Chichewa, translating as she went along, for Mercy Chipoka, a member of the National Smallholder Farmers' Association of Malawi (NASFAM), which had given her seed and advice. Mercy had grown pigeon peas for just 2 years on half an acre. She had learned that, unlike maize, this new crop did not need as much synthetic fertilizer, which was expensive and hard to get. But until the factsheet was read to her, she had no idea that pigeon pea captured nitrogen from the air and stored this in the soil. Extensionists tell farmers what to do, but often fail to explain the 'why'. Mercy seemed happy with this new piece of information.

Grace also learned a few things from Mercy. While pigeon pea was promoted as a monocrop to improve soil fertility, Mercy intercropped the pigeon pea with her maize. Farmers who eat maize every day are not willing to sacrifice their small piece of land to a new crop, but they will find a way to combine the two. And while Mercy did exactly this, she discovered that pigeon pea suppresses weeds. It helped her to save not only money, but also

Fig. 16.2. When farmers visualize their practices, such as mixed cropping patterns, it helps the video production team to better understand the topic.

labour. When experimenting with a new crop, farmers turn it around and look at it from many different angles, not just from the point of view of soil fertility improvement.

Half an hour later, Grace showed her factsheet to Shephard Bokosi, a young farmer from the same village. Shephard understood English and slowly read the factsheet. He had grown pigeon pea for 4 years on his one-acre (0.4-ha) field. He suggested adding some new information to the factsheet. The abundant leaves of the pigeon pea fall to the ground and decompose, so the following crop benefits from the improved soil. Pigeon pea can stay in the field for 3 years and give a harvest each year. It requires little labour.

The main section of the factsheet talked about how to plant pigeon pea as a monocrop, on ridges 75 cm apart. In Malawi tobacco is widely grown on ridges that are 1 m apart. When rotating with another crop, it is just too much work to plough the field, level it, and then make new ridges at a different distance. So when Shephard harvests his tobacco and plants maize and pigeon pea in the same field, he leaves in the ridges. He used black and red stones to show us how he plants both crops (Fig. 16.2). On one side of the ridge, he plants maize, one seed per hill, 25 cm apart, the way Sasakawa Global 2000 had taught him. At the other side of the ridge, he plants pigeon pea at 60 cm distance, the way NASFAM told him. Although both organizations train people on how to grow a monocrop, Shephard, like many other farmers, has taken parts of the advice and made these work under his conditions. He needs to get the most from his land and two crops yield more than one.

After Shephard rode away on his bicycle, Joseph Msaya, the local NASFAM field officer, arrived. Slightly excited, I explained to him the fascinating things we had just learned. Extensionists don't always like to have their recommendations critiqued, and predictably, Joseph rejected the idea: 'This will not work. By the time you harvest the maize, the pigeon pea has flowers and pods. When you cut off the maize stalks this will damage the pigeon pea.' It came down as a cold shower. I insisted we had to ask Shephard for more information. Fortunately, he had been chatting with a nearby farmer and so was still around, so we presented him with the problem. 'I no longer cut my maize stalks, but just harvest the cobs to avoid damaging the pigeon pea', Shephard replied simply. It was another great eye-opener. While extensionists aim to explain the broad rules of the game, it is the farmers who really play it, and know the strategies.*

Related video

Here is the video developed by the team in Malawi during the workshop.

NASFAM (2017) *Intercropping Maize with Pigeon Peas*. Hosted by Access Agriculture. 10 minutes.
www.accessagriculture.org/intercropping-maize-pigeon-peas

A version of the section above was previously published on www.agroinsight.com in 2014, by Paul Van Mele.

Aflatoxin Videos for Farmers

When developing videos for farmers, many things can go wrong. Yet most mistakes and frustrations can be avoided by proper research, planning and networking.

Projects that want to make farmer training videos do not always have a good idea of what farmers already know and do. The content of a video has to be shaped by the learning needs of the target audience, but in my experience, this is often given insufficient attention. It always pays to investigate farmers' knowledge and practices before you start filming, to avoid unpleasant surprises and mistakes (Van Mele, 2006; Bentley *et al.*, 2015).

In 2015, the International Crops Research Institute for the Semi-Arid Tropics (ICRISAT) asked Agro-Insight to explore what was needed to develop a series of farmer training videos on aflatoxins in groundnut. Aflatoxins are invisible, poisonous and cancer-causing chemicals produced by certain moulds. Since the 1960s, many West African countries have seen their groundnut production and trade dwindle due to aflatoxin contamination and stricter food safety standards imposed by European and North American markets.

During a scoping mission in September 2015, I met farmers in three different places in southern Mali who had received training on aflatoxins. Only by interacting with such empowered farmers can you discover what they have learned and gaps in their knowledge. A few knew a little about improved groundnut varieties that were resistant to aflatoxins or how to manage soils to suppress harmful fungi.

Mariam Coulibaly, president of a women's group in Wacoro, with about 70 women seed producers, was one of the few women I met who had carried out small experiments to assess the effect of compost on groundnut diseases. One of the roles of the video producer is to highlight such experiments and local innovations. For instance, I learned about using chilli powder against insect pests in seed storage, and using engine oil against termites, which damage groundnut pods in storage and increase the risk of aflatoxin contamination.

At times, triggering farmers to adopt good agricultural or post-harvest practices may be hard unless there are sufficient incentives. A Belgian NGO (VECO – *Vredeseilanden*, now Rikolto) had trained groundnut farmers in Uganda for several years on good practices to mitigate aflatoxin contamination. Despite extensive training, the farmers were reluctant to invest time and effort in removing mouldy kernels, because cleaned groundnuts fetched the same price as unsorted ones. A change in agriculture may demand social and economic innovations, which should be part of training videos.

Binta Coulibaly from Kolokani village in Mali helped us in this regard by stressing how women food processors have a social responsibility to protect their children and families from poisons in their food. Before converting the groundnuts into flour and baby food products, she carefully sorts out all the bad groundnuts, which she then turns into soap.

I thought this was a neat example of a practice adding an economic incentive to the tedious sorting. But after visiting farmers in Mali, I had a chance to interact with researchers and health specialists attending the roundtable of aflatoxin experts in Brussels on 'Building a multi-stakeholder approach to mitigate aflatoxin contamination of food and feed', organized by PAEPARD (Platform for African–European Partnership in Agricultural Research for Development). To my great surprise, I learned that aflatoxins are also transferred via the skin, so even soap can be dangerous! The next step would be to tell farmers about this, and to see if they can devise a better use for the damaged groundnuts. Sharing information between farmers and scientists can be a long-term dialogue.

For video topics that have such a global implication for human health, farmers' livelihoods and international trade, it pays to seek further international consensus on filming locations and the final content of the videos.*

Related video

Aflatoxins are a serious problem in stored food products, but there are simple ways to manage them.

Agro-Insight (2018) *Managing Aflatoxins in Maize before and during Harvest*. Hosted by Access Agriculture. 14 minutes.
www.accessagriculture.org/managing-aflatoxins-maize-and-during-harvest

A version of the section above was previously published on www.agroinsight.com in 2016, by Paul Van Mele.

Killing the Soil with Chemicals

Paul, Marcella and I were filming in 2022 in Quilcas, a village in Junín in the central Andes of Peru. A farmer and a former president of the community, Marcelo Tiza was spending the day with us. As we were admiring the mountain peaks and the green hillsides surrounding the community, we noticed that the steep slopes were divided up into a faint green checkerboard pattern, like a patchwork of abandoned fields. Then don Marcelo remarked offhand that all of that land had once been farmed, but that the soil had been destroyed by chemical fertilizer.

According to the community, these hillsides had always been cultivated, in a long rotation called 'turns', where they divided their high lands into several large fields, each about the same size. They would open one field the first year and divide it into family parcels of land to plant potatoes. The next year, they would open another big field for potatoes, and the first one, where they had already harvested potatoes, would be planted in other Andean tubers, or broad beans or some other crop. Then the land would rest for 5 years, until it became fertile again and people would plant potatoes again. (This is the open field system.)

Then in the 1970s, the people of Quilcas began to use chemical fertilizer to boost their potato yields. Some people could afford chemical fertilizer, and those who couldn't would apply sheep manure to their land. But after just 25 years of using chemical fertilizer, the communal land had been ruined. By 1999, community members noticed that even after they let the land rest for 5 years, it no longer recovered its fertility. It was missing its thick cover of vegetation. Gone were plants that indicated healthy soil, like *trébol de carretilla* (English: burr medic, *Medicago polymorpha* or *M. hispida*).

So, the people of Quilcas moved their communal land higher, from about 3800 m above sea level to nearly 4000 m. Having learned their lesson, the people prohibited the use of any chemical fertilizer or pesticides on these lands, and people who broke these rules could be fined or even lose their rights to the lands.

Since 2013, the community of Quilcas (in collaboration with the NGO Yanapai) has also learned to use a long rotation of fodder crops (grasses and legumes). For several years, they plant potato in rotation with other tubers, as well as with barley and oats. Then the land is rested for several years, by planting a cover of fodder crops, which enrich the soil. They have perfected this system in the individual lands near their homes, in the lower parts of the community (at about 3500 m above sea level). And now they are experimenting with planting fodder above the villages, in the soil spoiled by chemicals. The first yields have been good, and people are encouraged. Ecological farming may be able to restore soils that have been ruined by the intense use of chemicals.

Communities have the ability to notice change that has unfolded over several decades, at the level of whole landscapes, and to proactively experiment with ways of restoring the soil their lives depend on.*

◼️ Related video

Here is the video we filmed in Quilcas, Peru, on livestock and improved fodder.

Agro-Insight and Yanapai (2022) *Improved Pasture for Fertile Soil.* Hosted by Access Agriculture. 16 minutes.
www.accessagriculture.org/improved-pasture-fertile-soil

---

*A version of the section above was previously published on www.agroinsight.com in 2022, by Jeff Bentley.

Three Generations of Knowledge

> As a youth I planted a little and my grandparents told me nothing about these bioindicators. My potatoes had a lot of worms. I was discouraged and decided to seek another life.

So said don Miguel Ortega when we visited his farm in 2019 in Viloco village. Now in his mid-40s, don Miguel runs a prosperous organic farm in the Central Altiplano of Bolivia.

During his interview in front of the camera, don Miguel explained why he returned to his home village and picked up farming again:

> Because when you work in a company, coming on time, leaving on time, it is a form of slavery. So now that I work for myself, I am a free man.

In the meantime, don Miguel is one of the 70 *Yapuchiris*, expert farmers who share their knowledge with their peers and anyone who is interested in learning from nature and about healthy farming (Quispe *et al.*, 2008, Quispe *et al.*, 2018). But to become an expert farmer who can predict the weather based on observing plants, animals and insects has not been easy (see Choquetopa Rodríguez, 2021, for a longer description of these indicators, by one of the *Yapuchiris*). The elders in the village were not forthcoming with sharing their knowledge about natural indicators, as don Miguel explained:

When I asked the elders, they said 'in this way'. But you do not ask them just like that with the mouth empty. You have to give them a little soft drink. I managed it this way. I did not pick up a piece of paper at that moment. I held it in my mind. I held it in my mind and when I arrived home, I wrote it on paper. That is how I worked. By questioning. If we would pick up a sheet of paper and write they would not want to tell us everything.

Five days after meeting with don Miguel, we drove to the village of Ch'ojñapata, at an altitude of 4250 m. We interviewed Mery Mamani, who was in her early 20s. She ran a little shop where she sold soft drinks, beer and homemade cheese. Although we planned to interview her about an app that forecasts the weather, it soon became clear that this young woman had much more to tell us.

Full of energy, she guided us down the steep slopes to a valley behind her house. A pretty cactus with red flowers, called *sank'ayu* in the local Aymara language (*Echinopsis maximiliana*), is what she wants to show us. Mery said,

The app is great to tell us which day it will freeze or rain in the coming days, but this cactus tells us when is the best time to plant potatoes.

While Marcella films her in her little shop, Mery opens WhatsApp on her smartphone and shows photo after photo of various plants, mainly cactuses. All are bioindicators. Mery is clearly interested in making the right decisions on when to plant and do the other activities on her farm, and she cleverly combines knowledge from the past with modern forecasting. Youth like Mery who remain in the countryside, and who are interested in ancestral knowledge, can share those ideas and their observations with peers in other communities and other parts of the country. New communication devices can keep old knowledge alive.*

📹 Related videos

In the following video don Miguel and other Bolivian farmers show how they record and follow their seasonal weather forecasts.

Agro-Insight and PROSUCO (2019) *Recording the Weather*. Hosted by Access Agriculture. 14 minutes.
www.accessagriculture.org/recording-weather

A version of the section above was previously published on www.agroinsight.com in 2019, by Paul Van Mele.

References

Bentley, J.W., Chowdhury, A. and David, S. (2015) *Videos for Agricultural Extension. Note 6.* GFRAS Good Practice Note for Extension and Advisory Services. Available at: https://www.g-fras.org/en/good-practice-notes/6-video-for-agricultural-extension.html (accessed 16 October 2024).

Bentley, J.W., Van Mele, P., Harun-ar-Rashid, M. and Krupnik, T.J. (2016) Distributing and showing farmer learning videos in Bangladesh. *Journal of Agricultural Education and Extension* 22, 179–197.

Burns, S., Dittmer, K.M., Shelton, S. and Wollenberg, E. (2022) Global digital tool review for agroecological transitions. In: *Agroecological TRANSITIONS: Inclusive Digital Tools to Enable Climate-Informed Agroecological Transitions (ATDT)*. Alliance of Bioversity and CIAT, Cali, Colombia.

Choquetopa Rodríguez, B. (2021) *Indicadores Naturales para Pronosticar el Tiempo en el Sur de Oruro, Bolivia (Natural Indicators to Forecast the Weather in the South of Oruro, Bolivia)*. In: Bentley, J.W. (ed.). Agro-Insight and the McKnight Foundation, Cochabamba, Bolivia.

Liao, R. (2018) Exclusive: Alibaba Gets into Farming – Without Getting its Hands Dirty. TechInAsia. Available at: www.techinasia.com/alibaba-ai-et-brain-agriculture (accessed 16 October 2024).

Mann, L. and Iazzolino, G. (2021) From development state to corporate leviathan: Historicizing the infrastructural performativity of digital platforms within Kenyan agriculture. *Development and Change* 52, 829–854.

Microsoft News Center (2022) AGRA and Microsoft Extend Their Partnership to Support Digital Agricultural Transformation. Available at: https://news.microsoft.com/en-xm/2022/05/27/agra-and-microsoft-extend-their-partnership-to-support-digital-agricultural-transformation (accessed 16 October 2024).

Pascal, E., Canfield, M. and Fent, A. (2023) *Corporate-Led Climate Adaptation: How the Gates Foundation, Microsoft, and AGRA are Enabling the Digital Capture of African Food Systems*. AGRA Watch Report. Available at: https://cagj.org/wp-content/uploads/Digital-Ag-Report_Final.pdf (accessed 16 October 2024).

Quispe, M., Baldiviezo, E. and Laura, S. (2008) *Yapuchiris: Ofertantes Locales de Servicios de Asistencia Técnica (Yapuchiris: Local Technical Assistance Service Providers)*. PROSUCO, Cooperación Suiza, La Paz, Bolivia.

Quispe, M., Laura, S. and Baldiviezo, E. (2018) *Yapuchiris: Un Legado para Afrontar los Impactos del Cambio Climático (Yapuchiris: A Legacy to Face the Impacts of Climate Change)*. PROSUCO, Cooperación Suiza, La Paz, Bolivia.

Richards, P. (1989) Agriculture as a performance. In: Chambers, R., Pacey, A. and Thrupp, L.A. (eds) *Farmer First: Farmer Innovation and Agricultural Research*. Intermediate Technology Publications, London, pp. 39–43.

Scott, J.C. (2017) *Against the Grain: A Deep History of the Earliest States*. Yale University Press, New Haven, Connecticut.

Stone, G.D. (2022) Surveillance agriculture and peasant autonomy. *Journal of Agrarian Change* 22, 608–631.

Van Mele, P. (2006) Zooming-in, zooming-out: A novel method to scale up local innovations and sustainable technologies. *International Journal of Agricultural Sustainability* 4, 131–142.

Van Mele, P., Bentley, J.W., Harun-ar-Rashid, M., Okry, F. and Mourik, T. (2016) Letting information flow: Distributing farmer training videos through existing networks. *Indian Journal of Ecology* 43, 545–551.

Verdonk, T. (2019) Planting the seeds of market power: Digital agriculture, farmers' autonomy, and the role of competition policy. In: Reins, L. (ed.) *RegulatingNew Technologies in Uncertain Times*, Vol. 32. Information Technology and Law Series, T.M.C. Asser Press, The Hague, Netherlands.

17 Merging Scientific and Local Knowledge

Abstract

The integration of scientific and local knowledge can address various agricultural challenges. For example, by counting soil organisms like earthworms, farmers can assess soil health, understand the impact of farming practices and improve soil fertility by adding organic matter and reducing tillage. Diverse insect populations, nurtured by flowering plants, can naturally control pests and improve crop yields. Rotational grazing and fodder production systems, developed through farmer-led research, have restored degraded pastures and increased livestock productivity in the Andes. Bolivian farmers and researchers collaboratively developed low-cost, portable solar dryers to reduce aflatoxin contamination in groundnuts, emphasizing the importance of farmer innovation in agricultural technology. Farmers in India create cost-effective animal feed, but careful ingredient selection is crucial, especially for fish, where coconut oil cake can be lethal. The wealth of farmers' knowledge of African crops like sorghum and millet helped plant breeders develop climate-resilient alternatives to maize. Combined with local language videos on agroecological farming practices, the new varieties boosted the revival of these traditional crops.

Seeing the Life in the Soil

Earlier, we wrote about the importance of soil organic matter and soil life to support thriving, sustainable food production (see Chapter 6, this volume). Soils that have many living organisms hold more carbon and nutrients and can better absorb and retain rainwater, all of which are crucial in these times of a disturbed climate.

But measuring life in soils can be a time-consuming activity, depending on what one wants to measure. While bacteria and fungi cannot be seen by the naked eye, ants, grubs and earthworms can.

In one of the training videos that we filmed in Bolivia in 2023, Eliseo Mamani from the PROINPA Foundation, an agricultural research-and-development organization, shows how farmers can measure the visible soil organisms. Using a standardized method to measure soil life is important if you want to evaluate how certain farming practices have an effect on the life of your soil.

One early morning, we picked up Ana Mamani and Rubén Chipana in Chiarumani, Patacamaya, on the Altiplano about 100 km south of La Paz. Rubén and Ana took us to a field that had been cultivated for several years and that had not received any organic fertilizer. These farmers have learned through collaborative research that there are more living things in some parts of the field, and fewer in other parts. To show that on the video, they took samples from three parts of the field.

With a spade, they removed a block of soil 20 cm wide, 20 cm long and 20 cm deep. They carefully put all this soil in a white bag and closed it tightly so that the living things would not escape, because the earthworms and other living things move quickly.

We then drove to another place, where they collected three more samples from a field that had received organic fertilizer and where organic vegetables were grown. All samples were put in blue bags, all nicely labelled.

Under the shade of a tree, some more farmers had gathered to start counting the living organisms. One handful of soil at a time, they emptied each bag onto a plastic tray. As they came across a living creature, they carefully picked it out and reported it to Eliseo, who took notes: how many earthworms, how many ants, how many termites, how many beetles, how many spiders and how many grubs.

After an hour, the results were added up and samples compared. Only the field that had received organic fertilizer was rich in earthworms. The farmers discussed the findings as a group and concluded: If your soil has few living things, you can bring your soil to life by adding animal manure or compost, by leaving crop residues in the field and not burning them. You can also improve soil life by ploughing less, as ploughing disturbs bacteria, fungi and animals that add fertility to the soil.

Earthworms can be counted and used as soil health bioindicators. When done in collaborative research with farmer groups this helps farmers understand how certain farming practices affect the health of their soil and the long-term sustainability of their farm.*

◼️ Related video

The video described in this section explains how extensionists and communities can collaborate to learn about the life in healthy soil.

Agro-Insight and PROINPA (2023) *Seeing the Life in the Soil.* Hosted by Access Agriculture. 9 minutes.
www.accessagriculture.org/seeing-life-soil

A version of the section above was previously published on www.agroinsight.com in 2023, by Paul Van Mele.

More Insects, Fewer Pests

It's one of the great secrets of ecology that few insect species are pests. Most insects help us, by pollinating our crops, by making honey or silk and by killing insect pests, either by hunting them or by parasitizing them (Wyckhuys *et al.*, 2019). I was in Ecuador in 2022 with Paul and Marcella from Agro-Insight, along with Ecuadorian colleagues Carmen Castillo, Mayra Coro and Diego Mina, to make a video on the insects that help us.

Our first stop was the home of Emma Román and her husband, Luis Plazarte, in Aláquez, a parish near the city of Latacunga in the central Andes. On a small field behind their house, Emma explained that all flowering plants (trees, ornamentals or crops) attract insects, which feed on the pollen and nectar in the flowers. She has seen many beneficial insects: the bee fly (Spanish: moscabeja, *Eristalis* spp. [Syrphidae]) and the hairy fly (family Tachinidae), beetles (like the ladybird beetle) and true bugs. She adds, 'And there is a new one, the soldier fly (Spanish: mosca sapito, *Hedriodiscus* spp.).'

I was puzzled about the new insect. Perhaps an introduced one? Then I realized that since doña Emma has received training in insect ecology from Mayra and Diego, and has planted more flowering plants, she has begun to notice more kinds of insects, which are also becoming more abundant because of the flowers she plants. For example, she planted a row of lantana flowers to mark the boundary of her field. On the ground nearby, she pointed out some tiny spiders which we had not even noticed. 'You can see this one is carrying her eggs with her', she said, pointing to a whitish spider the size of a grain of rice. The family's small field of oats is surrounded by *pullilli* shrubs (*Solanum smithii*), and other plants like *chilca* (*Baccharis latifolia*) and the Andean cherry (*Prunus serotina*), which are visited by pollinating insects and others attracted by the plants' flowers.

As doña Emma's farm becomes insect-friendly, she notices more beneficial insects. The larva of the bee fly hunts and eats small, soft-bodied insects. The hairy fly lays its eggs in other insects. The hairy fly larva hatches inside the victim, eating it from the inside out. That's why doña Emma has few pests, even as she has more insects.

For doña Emma, the big advantage is that she can produce maize, blackberries and several kinds of vegetables with no pesticides. She says this means that she has tastier food that is healthier for her and for her family. And the diverse flowers around her house give her a sense of tranquillity and harmony.

As doña Emma put it, 'We plant a variety of plants for all kinds of insects, so that all the birds come, and they help us to conserve this ecosystem … to teach our children that there are these good insects and birds.'*

Related video

The video mentioned in this section shows how to attract beneficial insects with flowering plants.

Agro-Insight, IRD and INIAF (2022) *Flowering Plants Attract the Insects that Help Us*. Hosted by Access Agriculture. 15 minutes. www.accessagriculture.org/flowering-plants-attract-insects-help-us

*A version of the section above was previously published on www.agroinsight.com in 2022, by Jeff Bentley.

Moveable Pasture

Ideas for agricultural development are a bit like fads. They come and then fade away, for no apparent reason. For example, the local agricultural research committee (CIAL – *Comité de Investigación Agrícola Local*; Ashby, 2000) is a perfectly functional method for engaging farmers and researchers in joint experiments, which has been largely ignored in recent years. Google Scholar shows almost no mentions of the method since 2006. But where the CIAL has survived, it is still functional. The CIAL has led to lots of innovation in the community of Cordillera Blanca in the Peruvian Andes, where this committee continues to function after more than 20 years.

Every functional innovation we saw in the community seemed to be related to the CIAL. For example, Paul, Marcella and I met community member Trinidad León while she was herding her sheep home through the *bofedales*, the high Andean wetlands.

We found a place to get out of the wind behind doña Trinidad's stone cottage, where she explained that 30 years ago, overgrazing was a problem in the community (Fig. 17.1). Back then, there was no grass like we saw now. This surprised me, because this rocky pasture at 4000 m above sea level was thick with native needle grass when we saw it. Rotational grazing, moving the animals to let the pasture rest, had allowed this meadow to recover.

Rotational grazing is just one of the ideas that the CIAL and the community have experimented with over the years, working with different extensionists from the Mountain Institute, an NGO.

Doña Trinidad was not a member of the CIAL, but her husband was, and she knew well what the committee researched. Doña Trinidad explained how an agronomist named Doris Chávez worked with the community for several years, starting in about 2013, to discuss ways to improve pasture.

Previously, the couple would move their corral periodically and allow it to seed itself in native pasture. Through their interaction with the CIAL, they saw the opportunity to use the corral as a place to grow fodder, not just to allow pasture to grow naturally. At planting time, they plough the

Fig. 17.1. Marcella Vrolijks and Jeff Bentley interview Trinidad León as part of a video on rotational grazing.

old corral and plant it with oats or barley, which they cut to feed to their animals. Later, the harvested barley patch grows into natural pasture, which the sheep graze. The following year, the land can be fenced within a corral again, to gather manure. So, there is a 3-year rotation: corral, oats and barley, pasture, before starting over again with the moveable corrals (Fig. 17.2).

The CIAL is a committee of farmers, men and women, who test new ideas and share the results with their community. The farmers themselves adapt the ideas, and from what we saw, they can be successful. The oat and barley field is a healthy, emerald-green patch growing on the site of last year's corral. Doña Trinidad takes a sickle and cuts an armful to feed to her cattle later that afternoon.

Agroecology, with its emphasis on co-construction of knowledge, is now gaining importance across the world. Researchers today might take inspiration from the CIAL, as a way to stimulate community research, especially for agroecology.*

Fig. 17.2. Rotations of corrals, grains and pasture in the high Peruvian Andes.

📹 **Related video**

This is the video on rotational grazing, filmed in Peru, that is described in this section.

Agro-Insight and Mountain Institute (2022) *Rotational Grazing*. Hosted by Access Agriculture. 15 minutes.
www.accessagriculture.org/rotational-grazing

A version of the section above was previously published on www.agroinsight.com in 2022, by Jeff Bentley.

Making a Lighter Dryer

Fundación Valles, an NGO in Bolivia that does agricultural research and development, has developed a solar dryer to help prevent groundnuts from developing the moulds that produce deadly aflatoxins. The prototype model had an A-shaped metal frame, raised off the ground, and was covered in a special type of light-yellow plastic sheeting known as agrofilm, able to withstand long exposure to sunlight. The dryer kept out water, and with air flowing in from the ends of the dryer, the groundnuts could dry even on rainy days.

In 2016, in Chuquisaca, *Fundación Valles* worked with farmers to develop cheaper versions of the dryer, making the A-shaped frames from wooden poles instead of metal, and began distributing large sheets of agrofilm, 2 m by 12 m, for which farmers paid US$14, half the original cost. *Fundación Valles* encouraged the farmers to continue adapting the original design of the dryer. In May 2018, I visited some of these farmers, together with agronomists Walter Fuentes and Rolando Rejas of *Fundación Valles*, to find out what had happened.

When Augusto Cuba, in Achiras, received the agrofilm from *Fundación Valles* in 2016, he did not put it to immediate use. The weather was dry for several harvests, but on the rainy days during the groundnut harvest in May 2018, don Augusto put the agrofilm to the test. He took a plastic tarp to his field and laid it on the ground. He covered it with freshly harvested groundnuts, cut the agrofilm in half, and then placed the 6-m length on top.

Don Augusto ignored the basic design of the dryer. He didn't want to go to all of the trouble of cutting poles and building the raised platform. His design was much simpler, and portable: as he worked in the field, he could remove the agrofilm when the sun came out, and put it back when it started to drizzle again. The main disadvantage, however, was that the air did not flow over the covered nuts; humidity could build up, allowing mould to develop.

The original tent-like dryer has several limitations. It is expensive, and as don Augusto pointed out to us, it is a lot of work to make one from wood. At harvest, groundnuts are heavy with moisture. The pods lose about half their weight when dried. So, farmers dry their groundnuts in the field, and sleep there for several nights to protect the harvest from hungry animals. A solar dryer must be carried to the fields, yet these may be up to an hour's walk from home and involve climbing up and down steep slopes. Farmers who are using the original solar dryer, as designed by *Fundación Valles*, are those who have their fields close to home. Yet even taking a simple tarp to the harvesting site would be an improvement over drying the pods on the bare ground.

Later I had a chance to discuss don Augusto's method for drying groundnuts with Miguel Florido, an agronomist with *Fundación Valles*, and with Mario Arázola, the leader of the Association of Peanut Producers (APROMANI – *Asociación de Productores de Maní*), a groundnut farmers' association. They were concerned that don Augusto's design would trap in too much moisture, especially if it was misty all day and the farmer didn't have a chance to remove the agrofilm. We agreed that a dryer had to have a few simple agronomic criteria: it had to keep out the rain, keep the groundnuts off the ground and let air flow through.

After discussing don Augusto's case, we agreed that a dryer also had to meet some of the farmers' criteria: it has to be cheap, portable and able to handle large volumes of groundnuts, while keeping them out of the rain.

Aflatoxin contamination is a serious problem worldwide, and while it can be addressed, inventing a simple technology is hard work. Researchers start with a problem and some ideas to solve it, like air flow and keeping groundnuts dry. But it is only after offering farmers a prototype that researchers can see the farmers' demands. For example, designing a stationary dryer helps researchers to see that farmers need a portable one. Making and using a small dryer in the field highlights the need for a larger one. These types of demands only emerge over time, as in having a long, slow conversation, but one that is worth having.*

Related video

Later, in 2022, we returned to Achiras and filmed a video about how smallholders can manage aflatoxins.

Agro-Insight and *Fundación Valles* (2022) *Managing Aflatoxins in Groundnuts during Drying and Storage*. Hosted by Access Agriculture. 16 minutes.
www.accessagriculture.org/managing-aflatoxins-groundnuts-during-drying-and-storage

*A version of the section above was previously published on www.agroinsight.com in 2018, by Jeff Bentley.

Cake for Fish? Hold the Coconut, Please

A good farmer training video inspires farmers to modify practices, for example, replacing an ingredient of a locally made animal feed. But when changing ingredients, one has to know a lot about them, as we learned while teaching a video production workshop with Access Agriculture in Tamil Nadu, southern India, in 2016.

Explaining the principles behind a certain technology or why something is done in a particular way helps farmers to better understand the technology and to try it out with whatever resources they have at hand. The different examples shown in a video help to give farmers more ideas to work with.

In Tamil Nadu, one group of trainees was making a video on homemade animal feed, which only costs half as much as concentrated feed that one can buy in a shop (Figs 17.3 and 17.4).

By interacting with various farmers, the trainees learned quite a few things. While shops sell specific feed for different animals, farmers make a base mix of grains, pulses and oil cakes that they use to feed all their animals and fish. This saves the farmers time, while allowing them to still tailor the feed for each species of livestock. Depending on whether it is for cattle, goats, poultry or fish, they will then add some extra ingredients, like dried fish (if the feed is for fish or poultry).

The trainees also learned that when you want to use a base mix for fish, you need to consider a few things. Farmers rear up to six different species of

Fig. 17.3. Learning from farmers continues throughout the filming.

fish. Two species are surface feeders, two feed in the middle layer, and two species are bottom feeders. As you want the feed to be eaten by all fish, the mix should be milled to a coarse flour. When ground too fine, the feed will float and be available to the surface feeders only.

One other thing the team of trainees learned was that for fish, you can use groundnut oil cake or cotton seed oil cake, but you should never use coconut oil cake (which is readily available and cheap in coastal India). Because, if coconut oil cake is used in the base mix, 2 days after feeding the fish, an oily film will develop, blocking the pond from sunlight and oxygen and slowly killing the fish. The household can still use coconut oil cake in base feeds intended for livestock.

Clearly, oil cakes are not all the same and not all are interchangeable.

Good farmer training videos should present a range of different options and locally available resources, but they should also warn farmers of any possible risks. Videos for farmers should always say why an option will (or won't work), as in this case: don't feed coconut to your fish or the oil will block the sunlight and oxygen in the water and kill them!*

Fig. 17.4. Farmers at ease more readily share their knowledge.

📹 **Related video**

Here is the video described in this section about making livestock feed.

AIS, MSSRF and WOTR (2016) *Preparing Low-Cost Concentrate Feed*. Hosted by Access Agriculture. 10 minutes.
www.accessagriculture.org/preparing-low-cost-concentrate-feed

---

A version of the section above was previously published on www.agroinsight.com in 2016, by Paul Van Mele.

Sorghum and Millets on the Rise

For decades, various international aid agencies have pushed Africa towards adopting maize as the hunger-saving technical solution, with traditional crops such as sorghum and pearl millet only receiving a fraction of the support. But climate change is forcing donors and governments to rethink their food security strategies. Over the past two decades, research in Mali has revealed the importance of research and communication to help improve traditional crops and to support farmers as they cope with climate change.

While maize was first domesticated some 7000 years ago in Mexico, sorghum and pearl millet have their origin in Africa. Sorghum domestication started in Ethiopia and sub-Saharan Africa some 5000–6000 years ago. Through farmer selection, numerous improved sorghum types were developed, which then spread via trade routes into other regions of Africa and India. Domestication of pearl millet started only around 2500 BC, in eastern Mali, and spread rapidly to other countries through pastoralists, spurred by the increasing desiccation of the Sahara Desert at the time (Dillon *et al.*, 2007).

The rich genetic diversity of these traditional African crops and the wealth of farmers' knowledge have formed the basis of crop improvement programmes. In West Africa, a handful of devoted sorghum and millet breeders, Drs Eva and Fred Weltzien-Rattunde, Bettina Haussmann and Kirsten vom Brocke, in close collaboration with partners, were able to develop improved sorghum and millet varieties by working with local germplasm. The new varieties cope better with pests and diseases, as well as with rainy seasons that are becoming shorter and more unpredictable.

But these breeders, then working for the International Crops Research Institute for the Semi-Arid Tropics (ICRISAT), did not limit their efforts to participatory plant breeding alone; they also invested heavily in supporting farmer cooperatives to become seed producers and sellers. Some of these examples were captured in a chapter written by Daniel Dalohoun as part of the book *African Seed Enterprises* that Jeff and I edited with Robert Guéi from the FAO (Dalohoun *et al.*, 2010).

Farmers across Africa are keen to learn how to better conserve, produce and market seed of their traditional crops. While we were making a video on farmers' rights to seed at a seed fair in Malawi, farmers eagerly exchanged traditional sorghum and millet varieties with each other (see the section 'Richness in Diversity' in Chapter 5, this volume). As the government had so far focused only on maize as a food security crop, some communities lost certain traditional sorghum and millet varieties, but seed fairs and community seed banks helped them to again access these varieties. In addition to seed, farmers also want new knowledge about farming practices. Mr Lovemore Tachokera, a farmer from southern Malawi who attended a seed fair in the north, told me:

> The one thing I will make sure to tell my fellow farmers back home regarding conservation of indigenous crops is that we should also practise new farming technologies, even on the indigenous crops.

And right he was. Treasuring and improving traditional crops is important, but that alone is insufficient to cope with climate change; good agricultural adaptation strategies also matter. As part of his PhD work in Benin, Gérard Zoundji investigated how a series of farmer training videos on weed and soil management helped farmers in Mali to use agroecological practices (Zoundji *et al.*, 2018).

The differences in impact between video villages (where farmers had watched the videos) versus control villages (where no videos were shown) were significant. In the video villages, 99% of the farmers practised crop rotation combined with intercropping; 94% diversified their crops; and 51% used zaï pits. These figures were lower in the non-video villages, just 51% for crop rotation plus intercropping, 52% for crop diversification and 0% used zaï pits.

Zoundji also found that after watching the videos, farmers started demanding improved cereal seed. After watching the striga videos, some women's groups in the villages of Daga and Sirakélé became seed dealers in their village. Sorghum, millet and maize yields in the video villages increased by 14%, 30% and 15%, respectively, when compared to non-video villages (Zoundji et al., 2018).

While maize crops are increasingly failing in parts of Africa due to climate change, the robustness of traditional African cereal crops contributes to their renewed appeal for African farmers. The improved cultivation of traditional, drought-resistant crops, benefiting from research and training on improved cropping practices, will enable farmers to adapt to a harsher and more variable climate.*

Related video

To develop new varieties, researchers collaborate intensively with men and women farmers.

Agro-Insight (2016) *Succeed with Seeds*. Hosted by Access Agriculture. 11 minutes.
www.accessagriculture.org/succeed-seeds

A version of the section above was previously published on www.agroinsight.com in 2017, by Paul Van Mele.

References

Ashby, J.A. (2000) *Investing in Farmers as Researchers: Experience with Local Agricultural Research Committees in Latin America*. CIAT, Cali, Colombia.

Dalohoun, D.N., Van Mele, P., Weltzien, E., Diallo, D., Guindo, H. *et al.* (2010) Mali: When governments give entrepreneurs room to grow. In: Van Mele, P., Bentley, J.W. and Guéi, R. (eds) *African Seed Enterprises: Sowing the Seeds of Food Security*. CAB International, Wallingford, UK, pp. 65–88.

Dillon, S.L., Shapter, F.M., Henry, R.J., Cordeiro, G., Izquierdo, L. *et al.* (2007) Domestication to crop improvement: Genetic resources for *Sorghum* and *Saccharum* (Andropogoneae). *Annals of Botany* 100, 975–989.

Wyckhuys, K.A.G., Heong, K.L., Sanchez-Bayo, F., Bianchi, F.J.J.A., Lundgren, J.G. *et al.* (2019) Ecological illiteracy can deepen farmers' pesticide dependency. *Environmental Research Letters* 14, 093004.

Zoundji, G.C., Okry, F., Vodouhê, S.D. and Bentley, J.W. (2018) Towards sustainable vegetable growing with farmer learning videos in Benin. *International Journal of Agricultural Sustainability* 16, 54–63.

18 Speaking of Language

Abstract

This chapter explores the complex interplay of language, communication and media in agricultural development. Case studies from Mali, Malawi and Ecuador demonstrate how local languages, combined with visual storytelling and non-verbal cues, can effectively disseminate agricultural knowledge and foster community engagement. By highlighting the power of relatable content and authentic farmer voices, this chapter underscores the importance of culturally appropriate communication strategies in promoting agricultural innovation.

The Truth of Local Language

The many languages of Africa create niches for broadcasters like Gustave Dakouo, director of Radio Moutian in Tominian, Mali. *Moutian* means 'truth' in the Bomu language, spoken around Tominian.

Gustave runs his small commercial station with just three people, from a small building with a simple studio on the edge of the small town. And while Truth Radio may enjoy a monopoly among Bomu-speakers (between 100,000 and 200,000 people), one of the problems is finding enough content in the language to play on the air. The station broadcasts from 8 to 11 AM and again from 6 to 11 PM, except on weekends when they start at 4 PM. That's 8 or 10 h of airtime a day, and it all needs to be filled with programming.

Then in 2012, Gustave received copies of the *Fighting Striga* videos, which were published in Bomu and several other local languages. The videos gave background information and practical, affordable ideas for beating the striga weed. Gustave would play the soundtrack of one of the videos at the appropriate time of year (e.g. videos about planting just before the planting season), so people found the advice timely. His listening public reacted warmly. The area is heavily rural, where people grow sorghum and millet for a living, and striga, the devil weed, was strangling their crops.

Farmers began calling into the station, asking questions about the programmes. Gustave was a journalist, not an agricultural expert, so he asked

© Jeffery W. Bentley and Paul Van Mele 2025. *Agroecology in Practice: From Local Initiatives to Global Scaling Through Video* (J.W. Bentley and P. Van Mele)
DOI: 10.1079/9781800628793.0018

for help from Pierre Théra, an experienced farmer in Tominian. Pierre was also the head of the respected Union of Farmers of the District of Tominian (UACT – *Union des Agriculteurs du Cercle de Tominian*). Pierre had worked on striga for a long time, in collaboration with agronomists, and he knew the videos well. So, Gustave organized radio shows, where he would play the striga soundtracks and Pierre would come to the studio and answer the farmers' questions as they called in.

Making a call on a cell phone costs money, and if farmers are willing to ring up and ask questions, it means they are paying for information with their own money. Farmers also came to the station and asked Gustave for copies of the DVD, so they could watch the videos in their home villages. Fortunately, he had some copies to give them.

While there is a niche for journalists who can serve languages with few speakers, the tricky part is generating hours and hours of content with a staff of two or three people. It's a hard job, but Gustave seems happy. He estimates that he has reached 50,000 people with information about striga, and broadcasting the soundtracks has given his station much more popularity. His listeners needed the information that he had.*

Related video

This is the first of the series of striga videos, now available in 40 languages.

Agro-Insight, ICRISAT and UACT (2016) *Integrated Approach against Striga.* Hosted by Access Agriculture. 8 minutes.
www.accessagriculture.org/integrated-approach-against-striga

A version of the section above was previously published on www.agroinsight.com in 2014, by Jeff Bentley.

Friends You Can Trust

Smallholder farmers get most of their new ideas from other farmers, and generally trust friends, neighbours and other peers more than outside technical experts (Cofré-Bravo *et al.*, 2019). The farmers' friends usually live nearby. But other than convenience, the friends are valued because they are trusted. What works for my friends might work for me.

We saw a new twist on this in 2016 in Malawi when Ronald Kondwani Udedi and I were interviewing farmers who had watched learning videos distributed by DJs: young entrepreneurs who sell entertainment videos.

Most of the videos had been made elsewhere (not in Malawi). The videos, on rice, striga (the parasitic weed) and chilli, had then been narrated in some of the local languages (Chichewa, Senna and Yao). When we spoke with smallholders in Malawi, they often called the farmers in the videos their 'friends', as we heard from Fadwick Matolo, in Ulolo village, near Phalombe

(Bentley *et al.*, 2016). The videos themselves do not say that the farmers are 'friends', and the Malawian farmers had received the videos cold – so to speak – with no extensionist to suggest that the folks on the screen were 'friends'. The Malawian farmers themselves had decided (each one independently of other farmers) that the people on the screen were their friends. At first, I found this puzzling.

For example, Hope Mazungwi, in Stolo village, near Mulanje, took the videos to a village cinema where the owner let him play some of them. Hope recalls, 'We saw that our friends are doing amazing things. The rice has big eyes.' Hope's friends, in this case, were farmers that he had never met, in faraway Mali.

Esme Stena, near Chombe, watched the videos at a friend's house and later told us, 'Our friends in the video, they keep rice seed in a clay pot. Does that mean that we should also keep our rice seed in a clay pot?' In this case, Esme's 'friends' were women farmers in Bangladesh.

I had earlier noticed that farmers in Uganda referred to the smallholders on the screen as 'our brothers and sisters' (Bentley *et al.*, 2013).

The learning videos are filmed with farmers in various countries, but are made to be translated into other languages and shown all over the world. After all, tropical smallholders are already watching entertainment movies from foreign countries; they can just as easily watch learning videos from elsewhere. These learning videos are well made, capturing the viewers' attention with music, engaging interviews, beautiful photography and relevant topics. The videos feature relaxed farmers speaking from the heart about practical ideas that really work. They are honest farmers, who are not acting, and they gain the trust of the audience. With trust comes friendship.*

◢█ Related video

This is the video from Bangladesh, with Esme's 'friends' on how to keep rice seed healthy by storing it in clay pots. The link is for the English version, but the video is also available in Bangla (from Bangladesh), Chichewa (from Malawi) and many other languages.

Agro-Insight, CABI, Countrywise Communication, IRRI, RDA and TMSS (2003) *Rice Seed Preservation*. Hosted by Access Agriculture. 7 minutes.
www.accessagriculture.org/rice-seed-preservation

---

*A version of the section above was previously published on www.agroinsight.com in 2016, by Jeff Bentley.

Watching Videos with Farmers

While making a farmer training video in Bolivia in early 2023, we showed the community one of our previous videos. This turned out to be a useful way to help farmers understand the format of the video and how it would be produced. It also added motivation for farmers to join.

When we arrive at the village of Chigani Alto, on the shores of Lake Titicaca, at an altitude of 4000 m, 18 members of the local farmer association have gathered at the small building that displays a signboard saying it is a research and training centre on biological inputs. Compared to the factory we visited earlier (see the section 'Commercializing Organic Inputs' in Chapter 14, this volume), the two small rooms do not immediately display any sophisticated equipment. Next door is the small community's well-kept, picturesque church.

The community members bring lunch. The style is a bit like a picnic, called *apthapi* in Aymara, mostly traditional dishes in their woven blankets. As they open up their blankets on the tables, we see boiled maize on the cob, boiled cassava and native potatoes, *chuño* (rehydrated, freeze-dried potatoes), pieces of plantain in the skin, and a *chuño* version of oca, an Andean tuber. Some people brought broad beans, either green or brown ones that had been dried on the plant. A few prepared some salsa of tomatoes with pieces of fresh cheese. Others made traditional pancakes from mashed potatoes with pieces of onion and some other vegetables. What a delightful way to strengthen ties between community members.

Just as we were finishing this culinary feast, it started to hail, and judging by the dark sky, it would not stop in the next hour. Jeff suggested that, while we waited for the weather to clear, we could watch a video in Aymara, the local language spoken in this part of the country. The hail on the corrugated sheets of the roof, however, drowned out the sound from his laptop.

Fernando Villca, the young president of the farmers' association, suggested that we all move to the small room of the training centre, as there they have an insulated roof and a speaker system. With a little help from Roly Cota, the extensionist from Promotion of Sustainability and of Shared Knowledge (PROSUCO – *Promoción de la Sustentabilidad y Conocimientos Compartidos*), they connected the laptop to the speaker via Bluetooth. Young people are clearly tech-savvy wherever we go.

From start to finish, the women and men were completely captivated by the video on living barriers. From the video, the farmers recognized the *t'ula*, a group of local shrubs that the farmers in the video multiply from seed, to plant them in hedgerows to protect their soil from wind erosion. After the video, our audience immediately started to discuss between themselves, and then they turned to Jeff: 'Show us another one.'

In the meantime, the weather cleared and Roly Cota urged us to start with the filming. He promised the farmers that he would screen more videos for them in Aymara on his next visit. There is nothing more powerful than clearly articulated demand from farmers.

When Roly asked who was willing to join the filming activities over the next few days, all the farmers volunteered and a few asked to be interviewed in front of the camera.

At the end of the second day, all farmers who happily participated wanted to know when the video would be on the internet. 'We cannot wait to become world famous', laughs Juana Martínez, who has seen the benefits of organic foliar sprays on her potatoes. She clearly wants to become a fervent ambassador of homemade biofertilizer.*

📹 Related videos

Here is the video the farmers in Chigani Alto watched.

Agro-Insight and PROINPA (2018) *Living Windbreaks to Protect the Soil*. Hosted by Access Agriculture. 15 minutes.
www.accessagriculture.org/living-windbreaks-protect-soil

Here is the video we filmed with the farmers in Chigani Alto.

Agro-Insight and PROSUCO (2023) *Making Enriched Biofertilizer*. Hosted by Access Agriculture. 16 minutes.
www.accessagriculture.org/making-enriched-biofertilizer

---

*A version of the section above was previously published on www.agroinsight.com in 2023, by Paul Van Mele.

Language or Dialect? It's Complicated

People who speak different dialects of the same language can understand each other. Unlike different languages, the dialects of those tongues are 'mutually intelligible'. Americans and the British understand each other (almost always), because the US and the UK speak dialects of the same English language.

However, it's complicated. The classic example is Danish, Norwegian and Swedish, which are all fairly similar, but for political reasons and national pride their governments use the schools and the media to maintain the uniqueness of these languages, which are often mutually intelligible (Shariatmadari, 2019).

Arabic is an example in the other direction. Spoken in some 20 countries with important differences between each nation, the Arab countries consider themselves speakers of one language, based on a shared tradition in classical Arabic literature and other ties (Shariatmadari, 2019).

Quechua is a native American language still spoken in the Andes, in Ecuador, Peru and Bolivia. Quechua has many distinct dialects, and some of them are mutually unintelligible, making them languages in their own right. The Dutch linguist Willem Adelaar (2004) classifies the varieties of Quechua into four main groups:

- Quechua I – spoken in central Peru;
- Quechua IIA – spoken in northern Peru;
- Quechua IIB – spoken in Ecuador (where it is called 'Kichwa'); and
- Quechua IIC – spoken in Bolivia and southern Peru.

In 2022, with Paul Van Mele and Marcella Vrolijks from Agro-Insight, we visited the province of Cotopaxi in the Andes of Ecuador, where the agronomists Diego Mina and Mayra Coro study the lupin bean (*Lupinus mutabilis*) with several communities. Diego and Mayra took us to a Kichwa-speaking community, Cuturiví Chico, where we got a chance to find out if the local people understood the (Bolivian) Quechua version of our video on lupin beans. During a meeting with the community, Diego and Mayra invited them to watch the video, explaining that it had been filmed in Bolivia.

As the Quechua version of the video played, I watched the audience for their reaction. They smiled in appreciation. After all, videos in Quechua or Kichwa are equally rare. The farmers were absorbed in the 15-minute video all the way to the end.

Afterwards, Diego asked if they understood it. One person said he understood half. Another said, 'More than half, maybe 60%.' Then Diego asked the crucial question, 'What was the video about?'

The villagers neatly summarized the video. Diseases of the lupin bean could be controlled by selecting the healthiest grains as seed, and burying the sick ones. But the video had also sparked their imaginations. One said that in a previous experience they had learned to sort healthy seed potatoes, and now that they had seen the same idea with lupin beans, they wondered if the seed of broad beans could also be sorted to produce a healthier crop.

Diego still felt that the farmers hadn't quite understood the video, so he showed the Spanish version. But this time the reaction was muted. People watched politely, but they seemed a bit bored and at the end there was no new discussion. They basically understood the video the first time.

Language and dialect are valid concepts, but 'mutual intelligibility' can be influenced by visual communication, enunciation and motivation. For example, in this video, carefully edited images showed people separating healthy and diseased lupin beans, which may have helped the audience to understand the main idea, even if some of the words were unfamiliar.

Clarity of the speech also counts; this video was narrated by professional broadcasters, native speakers of Quechua, so the soundtrack was well enunciated. Motivation also matters; if a topic is of interest, people will strain to understand it. Lupin beans are widely grown in Cuturiví Chico, and these farmers really wanted to know about managing the crop's diseases.

After this experience, we filmed four videos in Ecuador and sent the Spanish versions of the scripts to be professionally translated, in writing. When I read the scripts I was convinced that Ecuadorian Kichwa and Bolivian Quechua are different enough to be called two separate languages. On the Access Agriculture website, both Quechua and Kichwa are now listed.

Sometimes it is possible for an audience to understand a video in a language closely related to their own. This is because the video also communicates visually, with photography. The audience understands more if the words are carefully and distinctly pronounced, and if the listeners are motivated by a topic that interests them.*

📹 **Related video**

The lupin video, which we filmed in Bolivia and screened in Cuturiví Chico, is available in various languages, including Quechua.

Agro-Insight and PROINPA (2017) *Growing Lupin without Disease.* Hosted by Access Agriculture. 12 minutes.
www.accessagriculture.org/growing-lupin-without-disease

---

*A version of the section above was previously published on www.agroinsight.com in 2022, by Jeff Bentley.

A Convincing Gesture

People use gestures intentionally to convey meaning, while many other hand movements are unconscious. Moving our hands helps us to grasp the right words. But human speech is also much more than words and hand gestures. Tone and volume of voice (screaming, whispering), facial expression, head movements (like nodding) and body language (slouching vs standing ramrod straight) all help to reinforce meaning and to convey emotion. We also make humming and clicking noises, which are sounds, but not speech. This non-verbal communication is convincing because it's natural. We can spot the difference; a phony smile is made with the lips only, while you use your whole face for a sincere one (Iverson and Goldin-Meadow, 1998; Corballis, 2012; Fröhlich *et al.*, 2019).

At Agro-Insight, when we make videos with farmers, we never tell them what to say. We ask them questions and film their answers, which we transcribe and translate into other languages. For example, if the farmer is speaking Arabic, we will use her voice in the Arabic version of the video, but we will dub over her voice for the English, French and other versions.

In these learning videos, the farmers' non-verbal communication is typical of unscripted, sincere speech. For example, in a video filmed in India, farmer Maran explained that he had a problem with the neighbours' turtles coming into his pond to eat the fish feed. As he said that, he moved his hands as if to suggest movement from one place to another. After hiring professional turtle catchers to remove the unwanted guests, everything was fine, an idea he reinforced by patting both hands downwards in a comforting gesture. The film crew didn't tell him to do that. Unless you watch the Tamil version of the video, you will hear a voice artist dubbing Mr Maran's words, but you can still tell that his gestures go with his narrative.

In the final cut of the video, we usually leave in some of the farmer's original voice before starting the voiceover. This lets the audience hear some of the emotion. For instance, in our video on feeding dairy goats, Teresia Muthumbi explains that when she gives her goats banana stems with sweet potato vines and a little grass, 'They give a lot of milk.' She is speaking from experience: you can hear the sound of authority in her voice, even if you don't understand Swahili.

In one video from Togo, farmer Filo Kodo tells how the maize harvest had increased a lot after rotating the maize with velvet bean (*Mucuna pruriens*). One neighbour even asked her what magic she had used. 'I told him it was with mucuna magic', she said, and you can see the smile in her eyes as well as on her lips.

As we say elsewhere, smallholders in Malawi called people on the farmer learning videos their 'friends', even though they had never met. Farmers in Uganda referred to their 'brothers and sisters' in West Africa, who they had only seen on the videos (Bentley *et al.*, 2013).

When people speak from the heart, their tone, gestures, expressions and body language convey conviction, even if the words themselves are translated into another language and spoken by another person. Non-verbal communication adds a richness, a sincerity that is hard to fake. This is one reason why realistic farmer-to-farmer training videos are a far richer experience than fully animated videos.*

Related videos

This section mentions three specific videos.

AIS, MSSRF and WOTR (2016) *Stocking Fingerlings in a Nursery Pond*. Hosted by Access Agriculture. 12 minutes.
www.accessagriculture.org/stocking-fingerlings-nursery-pond

Environmental Alert, DAES, KENFAP and Egerton University (2016) *Dairy Goat Feeding*. Hosted by Access Agriculture. 10 minutes.
www.accessagriculture.org/dairy-goat-feeding

Agro-Insight (2016) *Reviving Soils with Mucuna*. Hosted by Access Agriculture. 14 minutes.
www.accessagriculture.org/reviving-soils-mucuna

---

*A version of the section above was previously published on www.agroinsight.com in 2021, by Jeff Bentley.

References

Adelaar, W.F.H. (2004) *The Languages of the Andes*. Cambridge University Press, Cambridge, UK.

Bentley, J.W., Van Mele, P. and Musimami, G. (2013) The Mud on Their Legs – Farmer to Farmer Videos in Uganda. MEAS Case Study # 3. Available at: https://agroinsight.com/downloads/Articles-Agricultural-Extension/2013_AE5_MEAS-CS-Uganda-Farmer-to-Farmer-Videos-Uganda-BentleyJ-and%20PVanMele-July%202013.pdf (accessed 16 October 2024).

Bentley, J.W., Udedi, R.K. and Van Mele, P. (2016) Malawi DJs Distribute Videos to Farmers. Brussels, Belgium, Access Agriculture, for SDC. Available at: https://assets.access agriculture.org/s3fs-public/upload/files/Publications/Malawi%20DJs%20distribute%20 videos%20to%20farmers%20FINAL%20-%20Bentley%20et%20al%202016.pdf (accessed 16 October 2024).

Cofré-Bravo, G., Klerkx, L. and Engler, A. (2019) Combinations of bonding, bridging, and linking social capital for farm innovation: How farmers configure different support networks. *Journal of Rural Studies* 69, 53–64.

Corballis, M.C. (2012) How language evolved from manual gestures. *Gesture* 12, 200–226.

Fröhlich, M., Sievers, C., Townsend, S.W., Gruber, T. and van Schaik, C.P. (2019) Multimodal communication and language origins: Integrating gestures and vocalizations. *Biological Reviews* 94, 1809–1829.

Iverson, J.M. and Goldin-Meadow, S. (1998) Why people gesture when they speak. *Nature* 396, 228.

Shariatmadari, D. (2019) *Don't Believe a Word: The Surprising Truth About Language*. Weidenfeld and Nicolson, London.

19 Inspiring Use of Videos by Local Organizations

Abstract

Local organizations use farmer learning videos to promote agroecology in different ways. Charging a small fee for farmer learning videos in West Africa led to higher engagement than free distribution. Rural Malawians are rapidly adopting mobile technology, creating a vibrant local video culture that can be used to share agroecology and entertainment. Videos are used in school curricula in Kenya to teach kids good attitudes towards farming. Extensionists can combine videos with cost–benefit analysis to interest farmers in new food enterprises. A radio station in Ghana creates its own content by blending farmer learning videos with interviews. A university professor in Bolivia shows videos to inspire the next generation of farmers. A young Ugandan woman becomes a successful entrepreneur and leader, using farmer learning videos to empower other women.

Pay and Learn

Extensionists often give information away for free, but selling it may get you a more tuned-in audience. This is the conclusion of a paper published in *Experimental Agriculture* (Zoundji *et al.*, 2020).

Zoundji compared three groups of people in West Africa who had received DVDs of farmer learning videos. One video collection covered topics related to vegetable production and another showed how to manage the parasitic weed striga. The videos could be shown in multiple local languages, or in English or French (Zoundji *et al.*, 2020).

When NGOs in Benin gave the DVDs to organized groups of farmers, they tended to watch the videos, and they experimented with planting styles and other ideas shown in the videos. But some farmers who got DVDs for free did not show the videos to friends and neighbours, complaining that they needed fuel for their generators or other support (Zoundji *et al.*, 2020).

Audience appreciation improved when DVDs were shared by NGOs that were committed to the topic and the communities. In Mali, organizations that had taught striga management realized the importance of the weed, and arranged screenings of the videos in villages. Professional staff from the NGOs were on hand to answer people's questions after the show. The NGOs left copies of the DVD with local people who usually self-organized to watch the videos again later, to study the content. Farmers experimented keenly with the ideas they had learned, such as planting legumes between rows of cereal crops to control striga naturally.

But the big payoff came when farmers bought the DVDs cold, off-the-shelf in shops. Most only paid the equivalent of a dollar or two for the DVDs on vegetable production, but buying the information gave it value. All of these paying customers watched the videos and most of them showed the videos at home to friends and neighbours. They found the agricultural ideas useful; some bought drip irrigation equipment they had seen on screen, or creatively developed their own kit. Others learned to manage nematodes (microscopic worms) without chemical pesticides (Zoundji *et al.*, 2020).

Farmers who bought the DVDs also experimented with the digital technology used to show the videos. Nearly 15% bought DVD players to watch the videos. Some loaned the DVDs to their children at university, who copied the DVDs from the disk, converted them to a phone-friendly format (3GP) and then loaded the videos onto the mobile devices of friends and colleagues (Zoundji *et al.*, 2020).

Selling information draws a self-selected audience: interested people who will take the content seriously. Expert extensionists who appreciate the videos can also demonstrate their value by organizing video shows that respectfully engage with the communities and their leaders. But when DVDs are simply given away, even though they contain cinematic-quality videos on crucial topics, farmers may watch the videos and value them, or not. People who pay for information see its importance.*

Related videos

One of the striga videos was referenced in an earlier section, 'The Truth of Local Language' (see Chapter 18, this volume).

The other videos on striga can be seen at: www.accessagriculture.org/search/striga/all

---

A version of the section above was previously published on www.agroinsight.com in 2020, by Jeff Bentley.

Village Movies in Malawi

Young people in rural Malawi are getting into the digital age with the same enthusiasm as the rest of the world, but without all the same equipment. In Malawi, many people are using their cell phones as television sets (Bentley *et al.*, 2016).

Each small town, and even some of the villages, now has a computer person, called a 'DJ', who has a PC, often assembled in-country from imported parts. Most DJs are teenagers or 20-somethings who use freeware to convert videos into a format that phone memory cards can read. Farmers drop by the DJ's small shop (called a 'burning centre') and request Hollywood action films, or Nigerian or Bollywood movies. The films dubbed into Chichewa are a big hit, whether they follow the original story line or ad lib a new one. There are even Malawian movies made by artists like DJ Sau (*Only You*, a 90-minute love story), and standup comedy in Chichewa by Mr Jokes and others. Malawian gospel music videos are quite popular. Very few of the DJs have internet. But they visit each other and swap material.

The customers take their cell phone home and watch the movies in the evenings with their families on the screen of the phone (Fig. 19.1). If the room is very dark, several people can actually watch a movie on a screen the size

Fig. 19.1. Farmers in Malawi often watch entertainment and learning videos on their cheap feature phones. Photograph used with permission from Ronald Kondwani Udedi.

Fig. 19.2. Farmers in Malawi who are off the electrical grid can charge their phones in the nearby market towns, for a small fee.

of a matchbox. If the village lacks electricity, folks can have their phones charged while in town, at a shop that offers phone charging for a small fee (Fig. 19.2).

Few of the farmers have smartphones; they are simply watching films and videos on regular, inexpensive handsets.

Kids in the Global North make short video clips and share them via social media. The rural Malawians also have their own dense network, but largely disconnected from the internet. Some of the DJs make music videos and films with inexpensive cameras and swap their movies with each other. The films then circulate around to the other villagers, who watch them at home.

Many of the DJs give themselves cool names like Super DJ Andy T Man, who lives on a dirt road in the middle of nowhere, on the way to Lake Chilwa. Andy has made several music videos for a prosperous farm family, the Chigulumwas, who suspected that a neighbour was envious of their success and wanted to harm them with magic charms (Smith, 2019).

The song was composed and sung by Francis Masiye of Phalombe Town. Masiye praises Mr Chigulumwa, who grows good rice and sells it in the market. The video has shots of people hoeing a field, and the farmer paying them. The handsome couple and their teenage son dance on camera. The song says, 'I take care of my workers, and I work hard. And you are jealous of me and you use charms to try to hurt me.'

The aim of the song is to convince the jealous neighbour to stop trying to use magic against the Chigulumwa family. Everyone in the village has now seen the video and hopefully the jealous one got the message.

Villagers in the Global South are using their cell phones to watch dubbed foreign movies and local music videos. Some farmers are even making their own videos. This has been a rapid and recent change, suggesting that rural communities are now also ready to watch information videos about agriculture on their cell phones.*

◼ Related video

Ronald Kondwani Udedi discusses farmers who view videos in Malawi.

Malawi Broadcasting Corporation (2017) Interview with Kondwani Udedi. Hosted by EcoAgTube. 6 minutes.
www.ecoagtube.org/content/mbc-interview

A version of the section above was previously published on www.agroinsight.com in 2014, by Jeff Bentley.

Mix and Match

This book, especially Chapter 20, features farmers creatively adapting ideas after watching farmer learning videos. It should come as little surprise that agricultural extension people can also get inventive with new ways to show the videos.

In April 2017, I gave several organizations in Bolivia copies of a DVD with seven videos, each one with Quechua, Spanish and Aymara versions. Two of the videos were made with farmers in Bolivia, but the other five presented farmers from other countries.

María Omonte, an agronomist who was then the Bolivia director of the NGO World Neighbours, watched all seven of the videos. To my initial surprise, María also watched all of the videos in Spanish on the Access Agriculture video website. She checks the platform frequently to see if any new Spanish versions of videos have been added. María graduated from the prestigious agricultural university in Honduras, El Zamorano, and her training and natural curiosity have made her a keen life-long learner.

María and her team had been working for almost 3 years in six rural communities in Vila Vila, in the warm, semi-arid valleys of southern Cochabamba. In 2017, the team introduced the idea of organic fertilizer and Bordeaux mixture, a copper-based fungicide, as well as other similar products to control diseases of papaya, lemon and other crops. After seeing some practical demonstrations and receiving starter kits with the ingredients, some of the farmers tried the Bordeaux mix, but María felt that they needed more encouragement to keep making these mixtures.

So María creatively combined two videos. She took the Quechua version of a video on how to count costs and receipts from the DVD, and downloaded the Spanish version of *Turning Honey into Money*, which was not on the DVD, from the Access Agriculture website. She decided to use these two videos along with other information to make a unique training event for the six Quechua-speaking villages, a 5-hour drive from the city of Cochabamba.

I was there in 2018 in the community of Sik'imira at an evening meeting in the local school. The courtyard was full of high school students playing a furious game of football on the cement basketball court. María and her driver, Enrique Mancilla, set up their projector and within minutes, 25 farmers, over half women, had walked in and taken their seats.

María told the group that she had a video, dubbed into Quechua, that had been filmed in Mali, a country in Africa. From previous screenings María had learned that three details in the video were unfamiliar to farmers, so she explained that millet is 'a small grain', cowpea is 'a bean' and the money in the video is called 'the franc'.

By now the football game outside had ended and the teenagers were playing loud, pounding music. The grownups were too tolerant to ask the kids to turn the music down, but it did make the video a bit hard to hear. Still, people said they understood it, and they had no questions.

María used this as an opportunity to say: 'In the video on money we saw farmers and their facilitator adding up costs for different practices with millet and cowpea to see which one is more practical. Would you like to do the same with one of your crops?' The farmers suggested sweet potato.

It takes skill to walk through each step in the production of a crop and at the same time count the costs in front of an audience. Unfazed, María launched into the exercise in fluent Quechua. She started to struggle with the loud music still pounding next door, but eventually they turned it down enough for her to continue.

At the end, people looked at the results. 'We're not making much money', one said. 'That's sad', another added.

María used this as an entrée to discuss organic inputs to improve yields, then she asked the packed room – standing room only – if they would like to watch another video. It was 10 PM, past everyone's bedtime, but to my surprise the audience agreed.

Enrique and María put on the video *Turning Honey into Money*. She explained that this one was made in Kenya, also in Africa. By then, the music outside had mercifully stopped. The video played beautifully. The video was in Spanish, a language that some of the people spoke better than others. María told me that everyone understood the video, thanks to the clear images.

As the video ended, one man shouted out, 'Now we have the sweet taste of honey in our mouths!' Everyone laughed. None of the Bolivian farmers commented on the skin colour of the people in the videos, or their clothing. That is not an issue. Farmers in Nigeria (Bentley and Van Mele, 2011) and in Uganda (Bentley *et al.*, 2013) also like watching learning videos from other countries. María thinks that farmers are intrigued by seeing smallholders from far away.

After watching the honey video, María said that she could bring an expert beekeeper to help them get started raising bees. The farmers requested a meeting on Sunday morning. María and Enrique both agreed to give up their weekend to do that, delighted to have captured the farmers' imagination. Until now, the Sik'imira community had only ever wanted to meet at night. A Sunday morning meeting suggested that they were now taking the extension programme more seriously, helped by the warm response to the two videos.

A creative development professional, with access to a library of videos, can mix and match, combining a video on calculating farm costs with one on honey. Then she can explain more information to her audience, to make an exciting training event that local people find relevant.*

Related videos

Here is the video that Ing. María Omonte showed about honey.

NASFAM, NOGAMU, Egerton University and ATC/UNIDO (2016) *Turning Honey into Money*. Hosted by Access Agriculture. 11 minutes.
www.accessagriculture.org/turning-honey-money

The following video discusses finances on a small farm.

Philippine Permaculture Association, Alangilan National High School and NISARD (2024) *Record-Keeping for Integrated Farming*. 15 minutes.
www.accessagriculture.org/record-keeping-integrated-farming

A version of the section above was previously published on www.agroinsight.com in 2018, by Jeff Bentley.

Staying Grounded While on the Air in Ghana

It's a simple matter to play a soundtrack about farming on the radio. The tricky part is making sure that the programme connects with the audience, as I learned in 2021 from Gideon Kwame Sarkodie Osei at ADARS FM, a commercial station in Kintampo, a town in central Ghana.

Since 2010 Gideon had been pleased to be part of an effort by Farm Radio International (FRI) that supported radio stations in Ghana, including ADARS FM, to reach out to farmers. With encouragement from FRI, Gideon started a weekly show for farmers, where he plays Access Agriculture audio tracks. The programme is called *Akuafo Mo*, which means 'Thank You Farmers' in the Twi language. Before he started the show, Gideon (together with FRI) did a baseline study of the farmers in his audience. He found that they had more time on Monday evenings. Farm women do more work and have less time than men, but the women told Gideon that they were usually done with their chores by 8 PM, so that's when he airs *Akuafo Mo*, every Monday for an hour.

The show starts with recorded interviews, where farmers explain their own knowledge of a certain topic, like aflatoxin (a toxin produced by some fungi in stored food products). Aflatoxin is so important that Gideon had several episodes on this hidden danger. After the interviews, he plays an audio track, to share fresh ideas with his audience. Gideon has played Access Agriculture audios so often he can't remember how many he has played. 'It's a lot more than 50', he explains.

Gideon plays a portion of the audio in English, and then he stops to translate that part into Twi, the language of the Ashanti people. Every week there is a guest on the show, an extension agent who can discuss the topic and take questions from listeners who call in.

Gideon's experience with the radio programme inspired him to start listener groups, in coordination with FRI. Visiting listener communities, Gideon found that some did not have a radio set. So, with project support, he bought them one. 'We give them radio sets so they can come together weekly and listen to the programme', Gideon told me. He has 20 groups, each with 12 to 30 people. Five groups are only for women, especially in areas where males and females don't casually mingle. The other listener groups have men and women.

Gideon visits at least some of the groups every week. Because of these visits, he downloads videos as well as audio from Access Agriculture. Gideon explains:

> Sometimes I see if they have electricity, and I rent a projector, to show them the video they have heard on the air. This is my initiative, going the extra mile.

Some of the farmers are learning to sell their groundnuts, maize and other cereals as a group, netting them extra money and helping them to be self-sustaining.

Gideon is also a trainer for FRI. Before COVID-19, he would travel to other towns and cities in Ghana, meet other broadcasters, and go to the field with them to show them how to improve their interview skills and to craft their own radio shows. During the lockdown he trained broadcasters online.

Working with the farmer listening groups gives Gideon insights into farmers' needs and knowledge, making his radio programme so authentic that 60,000 people tune in. That experience gives Gideon the confidence to train other broadcasters all over Ghana.

When I was in Ghana a few years ago, I met excellent extension agents who told me how frustrated they were to be responsible for reaching 3000 farmers. It was impossible to have a quality interaction with all those farmers. However, there are ways to communicate a thoughtful message with a large audience, for example with a good radio show. There are many ways to play a video soundtrack on the air, but one of the best ways may be to blend it with a talk show or call-in radio. This can help extensionists to reach a much larger audience than they could do if they only met with farmers in person.*

> 📹 **Related videos**
>
> ---
>
> Access Agriculture hosts videos in Twi and in other languages of Ghana: Buli, Dagaari, Dagbani, Ewe, Frafra, Gonja, Hausa, Kabyé, Kusaal, Moba, Sisaala, Zarma and English.
>
> There are over 50 videos in Twi (at the time of publication):
> www.accessagriculture.org/search/all/tw

*A version of the section above was previously published on www.agroinsight.com in 2021, by Jeff Bentley.

Earthworms From India to Bolivia

In 2020, I met a young Bolivian journalist, Edson Rodríguez, who works on an environmental programme at the television channel TVU in Cochabamba, at the Public University of San Simón (UMSS – *Universidad Mayor de San Simón*). Edson helps to produce a show called *Granizo Blanco* (White Hail), a dramatic name in this part of the Andes, where hail can devastate crops in a moment. The show covers all environmental issues, not just agriculture. For example, the programme recently featured mud slides that have destroyed homes, and the impacts of a new metro train system in the valley.

When I first met Edson, I told him about the agroecological videos hosted on the Access Agriculture platform. Edson wondered if some of the videos might be suitable for his TV show. After watching some of the videos, he downloaded one on making compost with earthworms. The video was filmed in India, and it had recently been translated into Spanish. That made it possible to show a video from Maharashtra in Cochabamba. In spite of their linguistic differences, these parts of India and Bolivia have much in common, such as a semi-arid climate, and small farms that produce crop residues and other organic waste that can be turned into compost.

Edson asked me to take part in an episode of *Granizo Blanco* that included a short interview followed by a screening of the compost and earthworm video. He was curious to know why Access Agriculture promotes videos of farmers in one country to show to smallholders elsewhere. I said that the farmers may differ in their skin colour, clothing and hair styles, but they are working on similar problems. For example, farmers worldwide are struggling with crops contaminated with aflatoxins, poisons produced by fungi on improperly dried products like groundnuts and maize.

I told Edson that farmer learning videos filmed in Bolivia are being used elsewhere. Agro-Insight made a video, *Managing Aflatoxins in Groundnuts during Drying and Storage*, originally in Spanish, but it has since been translated into English, French and more than 30 African and Asian languages. The same aflatoxin occurs in Bolivia and in Burkina Faso, so African farmers can benefit from experience in South America. In this case, the video shows simple ways to reduce aflatoxins in food, using improved drying and storage techniques developed by Bolivian scientists and farmers.

'What other kinds of things can Bolivian farmers learn from their peers in other countries?' Edson asked me, as he realized that good ideas can flow in both directions. I explained that soil fertility is a problem in parts of Bolivia and elsewhere; Access Agriculture has videos on cover crops, compost, conservation agriculture and many other ways to improve the soil, all freely available for programmes such as *Granizo Blanco* to screen.

Many older people, especially government bureaucrats, feel that videos have to be made in each country and cannot be shared across borders. This closed vision makes little sense. The same civil servants happily organize and attend international conferences on agriculture and many other topics to share their own ideas across borders, and to 'harmonize' seed regulations (Bentley *et al.*, 2011). If government functionaries can gain insights from foreign peers, farmers should be able to do so as well.

Fortunately, younger people like Edson are able to see the importance of media, such as learning videos that enable farmers to share knowledge and experience cross-culturally. Smallholders can swap ideas and stimulate innovations as long as the soundtrack is translated into a language they understand. It costs much less to translate a video than to make one (Bentley *et al.*, 2015).*

◼ Related video

The Indian video on vermicompost that was shown on TV in Bolivia was available in 24 (mainly local) languages when we published this book.

WOTR (2018) *Making a Vermicompost Bed*. Hosted by Access Agriculture. 16 minutes. www.accessagriculture.org/making-vermicompost-bed

---

A version of the section above was previously published on www.agroinsight.com in 2020, by Jeff Bentley.

Videos to Teach Kids Good Attitudes

Kenyan schools recently changed to a Competency-Based Curriculum (CBC), moving away from memorizing facts and towards learning skills, knowledge and attitudes. One of the new topics is Information and Communications Technology (ICT), and another is Agriculture. Lawrence Njagi, the CEO of Mountain Top Educational Publishers, explained that the challenge was finding a way to integrate both subjects. He eventually decided that the best way was with videos from Access Agriculture.

In 2020, Mountain Top published a new textbook for fourth and fifth graders, to build students' confidence step-by-step. The textbook lists URLs for almost 20 videos on Access Agriculture, on gardening, legumes, pumpkins, small animals, innovative gardening and mulching. Teachers help students to pick a video topic, type in the URL and watch the video. Lawrence explained:

> They can watch the videos in either English or Kiswahili. It was great, because they could hear the voices of African people on the videos.

Ninety percent of the schools in Kenya are on the national electric grid, and 70% of those have access to Wi-Fi, including some schools in poor and remote areas. Watching the videos was 'an equalizing factor for those who could download', Lawrence says.

The students watch a video on, for example, making a vegetable seedbed. The textbook comes with a teachers' guide that explains how to lead the children in a project. The teacher organizes them in groups and the kids make a seedbed and plant kale in the school garden. The children also watch videos on how to make compost. Then they make the compost and fertilize their vegetables. The project lasts a whole term. The kids eat some of the vegetables, and on Parents' Day the proud students show their produce to the adults, who are allowed to buy some, teaching the students another valuable lesson: farming can bring in money.

This is important, because the Kenyan government is now encouraging young people to stay in the countryside. There are no more jobs in the cities. Young Kenyans have to employ themselves, and feed others while ensuring that Kenya is a food-sovereign nation.

Kenya's schools were closed for the COVID-19 pandemic, but they opened in October and November of 2020. During the closure, some schools and students tried to continue their studies with textbooks, educational TV and radio, and the internet. Some continued to watch Access Agriculture videos during the lockdown.

Kenya has 1.2 million pupils in each of grades four and five in 25,000 schools. Lawrence is optimistic that videos are important tools for education. 'We are equipping the children to produce food for themselves, and to sell.'*

Related video

Here is one of the videos featured in the Kenyan curriculum.

Agro-Insight, AMEDD, Countrywise Communication, FLASH, Fuma Gaskiya, ICRISAT, INRAN and UACT (2016) *Composting to Beat Striga*. Hosted by Access Agriculture. 10 minutes.
www.accessagriculture.org/composting-beat-striga

A version of the section above was previously published on www.agroinsight.com in 2021, by Jeff Bentley.

Teaching the Farmers of Tomorrow with Videos

Youth around the world are leaving agriculture, but many would stay on the farm if they had appropriate technologies and better social services, as Professor Alejandro Bonifacio explained to me.

Dr Bonifacio is from the rural Altiplano, the high plains of Bolivia. At 4000 m above sea level, this area has some of the highest farmland in the world. Bonifacio has a PhD in plant breeding, and besides directing an agricultural research station in Viacha, on the Altiplano, he teaches plant breeding at the public university in La Paz (Universidad Mayor de San Andrés).

The university attracts many rural youths. Every year, Bonifacio asks his new class of students to introduce themselves one by one and to tell where they come from, and to talk about their parents and their grandparents.

In 2021 about 20% of the students in Bonifacio's class were still living on the farm and taking their classes online. Another 50% were the children or grandchildren of farmers, but were now living in the city. Many of these agronomy students would be very interested in taking over their parents' farm, if not for a couple of problems.

One limitation is the lack of services in the rural areas: poor schools, bad roads, and no electricity or running water. While this is slowly improving, COVID-19 added a new twist, locking young people out of many of the places they liked to go to, and not just bars and restaurants. One advantage of city life is having greater access to medical attention, but this past year the students said it was as though the cities had no hospitals, because they were full of COVID-19 patients. Classes were all online, and so the countryside began to look like a nicer place to live than the city. Many students went home to their rural communities, where there was much more freedom of movement than in the city. (After the lockdown ended, most went back to the city.)

Dr Bonifacio told me that even when the youth do go home, they don't want to farm exactly like their parents did. The youngsters dislike back-breaking work with picks and shovels, but there is a lack of appropriate technology for young farmers, such as small, affordable machinery. Young farmers are also interested in exploiting emerging markets for differentiated produce, such as food that is free of pesticides. Organic agriculture also helps to save on production costs, as long as farmers have practical alternatives to agrochemicals.

Fortunately, there are videos on appropriate technologies, and Professor Bonifacio shows them in class. Today's youth have grown up with videos and find them convincing. Every year, Bonifacio organizes a forum for about 50 students on plant breeding and crop disease. He assigns the students three videos to watch, to discuss later in the forum. One of his favourites is *Growing Lupin without Disease*, which shows some organic methods for keeping the crop healthy. Bonifacio encourages the students to watch the video in Spanish and Quechua or Aymara. Many of the students speak Quechua or Aymara, or both, besides Spanish. Some feel that they are forgetting their native language. 'The videos help the students to learn technical terms, like the names of plant diseases, in their native languages', Bonifacio says.

During the COVID-19 lockdown, Professor Bonifacio moved his forum online and sent the students links to the videos. In the forum, some of the students said that while they were home, they could identify the symptoms of lupin disease, thanks to the video.

Bonifacio logs onto Access Agriculture from time to time to see which new videos have been posted in Spanish, to select some to show to his students so they can get some of the information they need to become the farmers of tomorrow.

Kids who grow up on small farms often go to university as a bridge to getting a decent job in the city. But others study agriculture and would return to farming if they had appropriate technology for family farming, and services like electricity and high-speed internet.*

Related video

In his classes, Professor Bonifacio shows the Spanish version of the video *Growing Lupin without Disease*.

Agro-Insight and PROINPA (2017) *Producir Tarwi sin Enfermedad*. Hosted by Access Agriculture. 12 minutes.
www.accessagriculture.org/es/producir-tarwi-sin-enfermedad

A version of the section above was previously published on www.agroinsight.com in 2021, by Jeff Bentley.

Giving Hope to Child Mothers

Teenage girls are vulnerable, and when they become pregnant, societies deal with them in different ways. In Uganda, they are called all sorts of names, such as a bad person, a disgrace to parents and even a prostitute. No one wants their children to associate with them because they are considered a bad influence. Parents often expel their daughters from the family and tell them that their life has come to an end. Rebecca Akullu experienced this at the age of 17. But Rebecca is not like any other girl.

After giving birth to her baby, she saved money to go to college, where she got a diploma in business studies in 2018. Rebecca soon got a job as an accountant at the Aryodi Bee Farm in Lira, northern Uganda, a region that has high youth unemployment and is still recovering from the violence unleashed by the Lord's Resistance Army, a rebel group. The farm director appreciated her work so much that he employed her.

'Over the years, I developed a real passion for bees', Rebecca says, 'and through hands-on training, I became an expert in beekeeping myself. Whenever I had a chance to visit farmers, I was shocked to see how they destroyed and polluted the environment with agrochemicals, so I became deeply convinced of the need to care for our environment.'

So, when Access Agriculture launched a call for young people to become Entrepreneurs for Rural Access (ERAs), Rebecca applied. The ERAs are farm advisors who use a solar-powered projector to screen farmer training videos as part of their business. After being selected as an ERA in 2021, she received the projector and training. At first she combined her ERA services with her job at the farm, but by the end of the year she resigned. Promoting her new business service required courage. Asked about her first marketing effort, Rebecca said she informed her community at church, at the end of Sunday service.

'I was really anxious the first time I had to screen videos to a group of 30 farmers. I wondered if the equipment would work, which video topics the farmers would ask for and whether I would be able to answer their questions afterwards', Rebecca recalls. Her anxiety soon evaporated. Farmers wanted to know what videos she had on maize, so she showed several, including the ones on the fall armyworm, a pest that destroys entire fields. Farmers learned how to monitor their maize to detect the pest early, and they started to control it with wood ash instead of toxic pesticides.

Rebecca was asked to organize bi-weekly shows for several months, and she continues to do this whenever asked. Having negotiated with the farm leader, each farmer pays 1000 Ugandan Shillings (0.25 Euros) per show, where they watch and discuss three to five videos in the local Luo language. Some of the videos are available in English only, so Rebecca translates them for the farmers. 'But collecting money from individual farmers and mobilizing them for each show is not easy', she says.

The videos impressed the farmers, and the ball started rolling. Juliette Atoo, a member of one of the farmer groups and primary school teacher in Akecoyere village, convinced her colleagues of the power of these videos, so Barapwo Primary School became Rebecca's second client, offering her another unique experience. Rebecca says:

> The children were so interested to learn and when I went back a month later, I was truly amazed to see how they had applied so many things in their school garden: the spacing of vegetables, the use of ash to protect their vegetable crops, compost making, and so on. The school was happy because they no longer needed to spend money on agrochemicals, and they could offer the children a healthy, organic lunch.

As she grew more confident, new contracts with other schools soon followed. For each client, Rebecca negotiates the price depending on the travel distance, accommodation and how many children watch the videos. Often five videos are screened per day for two consecutive days, earning her between 120,000 and 200,000 Ugandan Shillings (30–50 Euros). Schools will continue to be important clients, because the Ugandan government has made skill training compulsory. Besides home economics and computer skills, students can also choose agriculture, so all schools have a practical school farm and are potential clients.

While she continues to engage with schools, over time Rebecca has partly changed her strategy. She now no longer actively approaches farmer groups, but rather explores which NGOs work with farmers in the region and what projects they have or are about to start. Having searched the internet and done background research, it is easier to convince project staff of the value of her video-based advisory service.

As Rebecca, now the mother of four children, does not want to miss the opportunity to respond to the growing number of requests for her video screening service, she is currently training a man and a woman in their early 20s to strengthen her team.

Having never forgotten her own suffering as a young mother, and having experienced the opportunities offered by the farmer learning videos, Rebecca also decided to establish her own community-based organization: the Network for Women in Action, which she runs as a charity. Having impressed her parents, in 2019 they allowed her to set up a demonstration farm (Newa Api Green Farm) on family land, where she trains young girls and pregnant teenage school dropouts in artisan skills such as making paper bags, weaving baskets and making beehives from locally available materials.

Traditional beehives are made from tree trunks, clay pots or woven baskets smeared with cow dung that are hung in the trees. To collect the honey, farmers climb the trees and destroy the colonies. From one of the videos made in Kenya, the members of the association learned how to smoke out the bees, and not destroy them.

From another video, made in Nepal, *Making a Modern Beehive*, the women learned to make improved beehives in wooden boxes, which they construct for farmers upon order. From the video, they realized that they needed to make their bee boxes smaller. 'Because small colonies are unable to generate the right temperature within the large hives, we only had a success rate of 50%. Now we make our hives smaller, and eight out of ten hives are colonized successfully', says Rebecca.

Young women often have no land of their own, so members who want to can place their beehives on the demo farm.

> We also have a honey press. All members used to bring their honey to our farm. But from the video *Turning Honey into Money*, we learned that we can easily sieve the honey through a clean cloth after we have put the honey in the sun. So now, women can process the honey directly at their homes.

The bee business has become a symbol of healing. Farmers understand that their crops benefit from bees, so the young women beekeepers are appreciated for their service to the farming community. But also, parents who had expelled their pregnant daughter, embarrassed by societal judgement, begin to accept their entrepreneurial daughter again as she sends them cash and food.

'We even trained young women to harvest honey, which traditionally only men do. When people in a village see our young girls wearing a beekeeper's outfit and climbing trees, they are amazed. It sends out a powerful message to young girls that, even if you become a victim of early motherhood, there is always hope. Your life does not end', concludes Rebecca.

Stories from teams of Entrepreneurs for Rural Access (ERAs) from 15 countries across Africa and Asia are described in the book *Young Changemakers* (Van Mele *et al.*, 2024).*

Related videos

See these videos on beekeeping, mentioned in the above section.

Practical Action Nepal (2018) *Making a Modern Beehive*. Hosted by Access Agriculture. 15 minutes.
www.accessagriculture.org/making-modern-beehive

NASFAM, NOGAMU, Egerton University, ATC/UNIDO (2017) *Turning Honey into Money*. Hosted by Access Agriculture. 11 minutes.
www.accessagriculture.org/turning-honey-money

A version of the section above was previously published on www.agroinsight.com in 2023, by Paul Van Mele.

Videos to Encourage Agroecological Food Systems

Agrochemicals can be sold, but agroecological knowledge often has to be shared for free. In 2012, Access Agriculture (a non-profit) began to offer free videos on agroecology for farmers. A recent review of 244 digital tools found that Access Agriculture was one of only three that offered advice to smallholders on a wide range of agroecological principles, using exemplary extension features such as options in various languages (Burns *et al.*, 2022).

In 2021 we held an online survey of the users of Access Agriculture to find out how people were using and sharing the videos and other information. They could take the survey in English, French or Spanish, and 2976 people did so (Bentley *et al.*, 2022). Most of the respondents (83%) were living in Africa, where Access Agriculture started, suggesting that there is scope to expand in Latin America and Asia. Most survey takers were extensionists, educators (who show videos in class) and farmers themselves, who are increasingly getting online.

Access Agriculture makes an effort to feature female-friendly innovations and to film women farmers (as well as men). Still, 84% of the respondents were men. This is partly because women have less access to phones and to the internet (African Union Commission, 2021), but the videos do reach women. Many of the extensionists who were surveyed use the videos with organized groups of women farmers.

The survey asked how the videos had made a difference in farm families' lives. Answers were multiple choice, and more than one response was allowed. Choices were randomized so that each respondent saw them in a different order, so as not to favour the first items on the list. The top response, 'better yield', garnered almost 50% of the responses. This suggests that strengthening farmers' knowledge on agroecology through the videos can improve farmers' yields, an idea that is currently debated (Anderson *et al.*, 2021, p. 18).

The other frequent answers suggest that the videos promote productive, sustainable agriculture. 'Improved pests, disease and weed management', 'better soil health and soil fertility' and 'better produce' were all noted by over 40% of respondents. Only 1% thought that the videos had made no impact on farmers' lives.

Three-quarters (72%) of the farmers who download the videos also share them. Farmers would only do this if they found them useful. The survey estimated that since 2015, the videos reached 90 million people, mainly by mass media. That is partly because the videos are professionally filmed, and TV stations can request the broadcast-quality versions and play them on the air. Radio stations also broadcast the soundtracks, which are easily downloadable. Between 2012 and 2021, 4 million people were reached by local organizations, often screening videos in the villages.

Smartphones make it easy to share links to videos. Over half (51%) of the respondents shared the videos this way, reaching nearly 5000 (4927) organizations. By 2021, Access Agriculture had videos in 90 languages. However, only 55% of the survey respondents knew about these other language versions. As a result, by 2024, Access Agriculture had made local language versions easier to find online. In 2021, the Access Agriculture interface was only in three languages. Now it is in seven, as Arabic, Hindi, Bengali and Portuguese have joined English, French and Spanish. Access Agriculture has also begun to list the video title and written summary in the language of each version, not just in the languages of the interface. Now users can find videos by entering search words in languages like Kiswahili, Telugu and Quechua.

The farmers (and others) who took our survey are people who can afford the airtime to take an online survey. They are literate in English, French or Spanish, because they have had a formal education. But with time, smartphones will become less expensive to use. As today's youngest farmers mature, they will also bring more digital skills into the farming community. The next decade will make these videos even more accessible for farmers, extensionists and others, in ways we can scarcely imagine now. The Access Agriculture mobile app, downloadable from Google Play, is further enabling access to quality learning videos.

Agroecology relies on techniques such as crop rotation, organic fertilizer and natural enemies of plant pests. Many of these practices cannot be bought and sold. They depend on knowledge that can be conveyed online, by extensionists and in schools. Videos in many languages can effectively share agroecological knowledge and practices with farmers and students, for free, on the internet.*

Related video

As mentioned in this section, Access Agriculture hosts videos that feature female farmers to encourage more women to visit the platform and to become more interested in agroecology, as in this recent example.

Philippine Permaculture Association, Alangilan National High School and NISARD (2024) *Record-Keeping for Integrated Farming*. Hosted by Access Agriculture. 15 minutes. www.accessagriculture.org/record-keeping-integrated-farming

---

*A version of the section above was previously published on www.agroinsight.com in 2024, by Jeff Bentley.

References

African Union Commission (2021) *Africa's Development Dynamics 2021: Digital Transformation for Quality Jobs*. OECD Publishing.

Anderson, C.R., Bruil, J., Chappell, M.J., Kiss, C. and Pimbert, M.P. (2021) *Agroecology Now! Transformations Towards More Just and Sustainable Food Systems*. Palgrave Macmillan, Springer Nature, Cham, Switzerland.

Bentley, J.W. and Van Mele, P. (2011) Sharing ideas between cultures with videos. *International Journal of Agricultural Sustainability* 9, 258–263.

Bentley, J.W., Van Mele, P. and Musimami, G. (2013) The Mud on Their Legs – Farmer to Farmer Videos in Uganda. MEAS Case Study # 3. Available at: https://agroinsight.com/downloads/Articles-Agricultural-Extension/2013_AE5_MEAS-CS-Uganda-Farmer-to-Farmer-Videos-Uganda-BentleyJ-and%20PVanMele-July%202013.pdf (accessed 16 October 2024).

Bentley, J.W., Van Mele, P. and Reece, J.D. (2011) How seed works. In: Van Mele, P., Bentley, J.W. and Guéi, R. (eds) *African Seed Enterprises: Sowing the Seeds of Food Security*. CAB International, Wallingford, UK, pp. 8–24.

Bentley, J.W., Udedi, R.K. and Van Mele, P. (2016) *Malawi DJs Distribute Videos to Farmers*. Brussels, Belgium, Access Agriculture, for SDC. Available at: https://assets.accessagriculture.org/s3fs-public/upload/files/Publications/Malawi%20DJs%20distribute%20videos%20to%20farmers%20FINAL%20-%20Bentley%20et%20al%202016.pdf (accessed 16 October 2024).

Bentley, J.W., Chowdhury, A. and David, S. (2015) *Videos for Agricultural Extension. Note 6.* GFRAS Good Practice Note for Extension and Advisory Services. Available at: https://www.g-fras.org/en/good-practice-notes/6-video-for-agricultural-extension.html (accessed 16 October 2024).

Bentley, J.W., Van Mele, P., Chadare, F. and Chander, M. (2022) Videos on agroecology for a global audience of farmers: An online survey of Access Agriculture. *International Journal of Agricultural Sustainability* 20, 1100–1116.

Burns, S., Dittmer, K.M., Shelton, S. and Wollenberg, E. (2022) Global digital tool review for agroecological transitions. In: *Agroecological TRANSITIONS: Inclusive Digital Tools to Enable Climate-Informed Agroecological Transitions (ATDT)*. Alliance of Bioversity and CIAT, Cali, Colombia.

Smith, J.H. (2019) Witchcraft in Africa. In: Grinker, R.R., Lubkemann, S.C., Steiner, C.B. and Gonçalves, E. (eds) *A Companion to the Anthropology of Africa*. Wiley Online Library, pp. 63–79.

Van Mele, P., Mohapatra, S., Tabet, L. and Flao, B. (2024) *Young Changemakers: Scaling Agroecology Using Video in Africa and India*. Access Agriculture, Brussels, Belgium.

Zoundji, G.C., Okry, F., Vodouhê, S.D., Bentley, J.W. and Witteveen, L. (2020) Commercial channels vs free distribution and screening of learning videos: A case study from Benin and Mali. *Experimental Agriculture* 56, 544–560.

20 Encouraging Farmers to Experiment with New Ideas

Abstract

Videos are one of the excellent ways to share new ideas that encourage farmers to experiment. Learning videos empower Kenyan farmers and communities to adopt agroecological practices, revitalize local food systems and improve their livelihoods. A Cameroonian agricultural expert uses Access Agriculture videos to inspire farmers across Africa, sparking many local innovations through his television programme. Inspired by farmer-to-farmer videos, Malawian small-holder Lester Mpinda successfully expands his income by cultivating new crops and establishing a community-based chilli production initiative. Strong leadership and access to videos on lupin beans empower a Bolivian women's group to try new farming practices. An Ecuadorian farmer successfully adapts the biofertilizer techniques she learned from farmers in India, by means of a farmer-to-farmer video.

Scaling the Slow Food Movement in Kenya and Beyond

Eager to help rural people, Elphas Masanga studied agriculture and biotechnology at Bukura Agricultural College in Western Kenya, followed in 2014 by a 6-month practical course at the Kenya Institute of Organic Farming, and some short courses on permaculture and biodynamic farming.

Then Elphas began working for a local NGO, Seed Savers Network, where he learned the importance of biodiversity conservation, before joining Slow Food Kenya.

By 2019, Elphas had discovered Access Agriculture videos and was using them to train farmers. In 2021, Elphas was accepted into Access Agriculture's network of young Entrepreneurs for Rural Access (ERAs), and he received a solar-powered smart projector to screen videos in farm villages.

The videos *Farmers' Rights to Seed: Experiences from Guatemala* and *Farmers' Rights to Seed: Experiences from Malawi* convinced farmers that food sovereignty starts with becoming guardians of traditional crop varieties. 'When

DOI: 10.1079/9781800628793.0020

farmers see other farmers from Guatemala or Malawi, they are so excited. It gives them extra motivation and mileage', Elphas says.

Food Biodiversity and Agroecology is one of Slow Food Kenya's five strategic areas. 'We have screened many videos on soil management, pest and disease management, and marketing of agroecological farm products. The Access Agriculture videos have played a vital role in our work', says Elphas. In 2022 and 2023 he established 65 new community and school gardens and taught agroecology to nearly 300 schoolchildren, and more than 1440 adults, including 31% youth and 62% women.

As an ERA, Elphas is encouraged to use his smart projector to make money. While Slow Food Kenya has contracted Elphas various times to show videos in villages, one of his main clients is Participatory Ecological Land Use Management (PELUM) Kenya, a network of 65 civil society organizations. PELUM often requests Elphas to screen Access Agriculture videos during meetings and training of trainers' (TOT) sessions, for which he charges 3500 Kenyan Shillings (25 Euros).

When he leases the smart projector kit to other organizations, Elphas makes sure to be the one who operates the equipment to avoid anything breaking down. In 2022, he earned 68,000 Shillings (488 Euros) from leasing the kit.

Many farmers want their own copies of the videos, so Elphas also sells preloaded flash drives and DVDs for 50 Kenyan Shillings (0.36 Euros), and an extra 20 Shillings (0.14 Euros) for each video loaded.

Besides the videos from Guatemala and Malawi on farmers' rights to seed, Elphas has also screened two videos filmed in India, on *Community Seed Banks* and *Collecting Traditional Varieties*. By the end of 2023, this inspired the creation of 24 community seed banks in Kenya.

One seed bank was started by the Belacom women's group in Gilgil, in Nakuru county. These 15 women take pride in growing, selling and exchanging seed of many crops, such as Russian comfrey, kale, spinach, amaranth and cassava. The women sell their seed and agroecological produce at the weekly Gilgil Earth Market, part of the global network of Slow Food Earth Markets managed by the farmers, without middlemen. To strengthen the local organizations involved in Earth Markets, Elphas has screened videos about food marketing, including some filmed with farmers in Latin America on *How to Sell Ecological Food*, *Creating Agroecological Markets* and *A Participatory Guarantee System*.

Elphas starts each session by browsing the Access Agriculture video library on the projector and letting the farmers choose the videos to watch. He shows them the titles that are available in Kikuyu, a local language. Sometimes, he screens videos in Kiswahili. If the farmers want to see videos that are not yet translated into these languages, then Elphas shows the English version.

The 20 members of the Kahua-ini community garden group from Wanyororo watched the video *Good Microbes for Plants and Soil*, featuring farmers in India. Later, the garden group in Kenya started producing their own good microbes. Mungai Gathingu, who hosts a seed bank and is custodian of the community garden, used to grow French beans under contract farming on his half-acre (0.2-ha) field. At one point, the farm was producing almost nothing. By watching videos and putting what they learned into practice, Mungai and fellow members learned to improve their soil fertility, increase crop production, diversify their farms and earn good money by selling organic produce along with bottled mixtures of good microbes at the Slow Food Nakuru Earth Market.

After watching the video *Organic Biofertilizer in Liquid and Solid Form*, the 18 members of the Bee My Partner youth group in Njoro started a thriving business producing solid biofertilizer. This has earned them respect in their community and completely changed their attitude towards agriculture. They invested their earnings in beekeeping and fish ponds. They now also rear black soldier flies to feed the fish and poultry on their integrated farms.

In 2023, Pendo Internationale Zusammenarbeit, a non-profit organization of Kenyans living in Germany, engaged Elphas to screen Access Agriculture videos to schools in Nakuru county, for which he was paid 145,000 Kenyan Shillings (1040 Euros). Elphas trained hundreds of children in various schools.

The video *Teaching Agroecology in Schools* showed how a school in Peru taught the value of local culture and of a healthy lifestyle by including farming and traditional food in the school curriculum. 'After watching the video, the Kenyan schools began to use drawing, singing and poetry to teach the children about healthy food. The children began to see agriculture as a career rather than as a punishment', says Elphas.

As the coordinator of the Slow Food Youth Network in Kenya and the communication person for the Slow Food Youth Network in Africa, Elphas realizes that farmers need a different kind of advice:

> The government extension service in Kenya is decentralized and heavily supported by multinationals such as Syngenta. Therefore, many government extension workers only promote seed and agrochemicals from companies. Kenya needs more ERAs like myself and more smart projectors, to help farmers explore ecological farming and healthy food through Access Agriculture videos.*

▮ Related video

All of the videos that Elphas uses can be found on www.accessagriculture.org. The following is a video about Elphas and his work as an ERA.

Access Agriculture (2022) *Entrepreneurs for Rural Access (ERA): Elphas Masanga*. Hosted by EcoAgTube. 3 minutes.
www.ecoagtube.org/content/entrepreneurs-rural-access-era-elphas-masanga

---

*A version of the section above was previously published on www.agroinsight. com in 2024, by Paul Van Mele. More stories can be found in the book Young Changemakers (Van Mele et al., 2024).

A Greener Revolution in Africa

After settling in the USA in the 1990s, Isaac Zama would visit his native Cameroon almost every year, until war broke out in late 2016, and it became too dangerous to go home. About that same time a new satellite TV company, the Southern Cameroons Broadcasting Corporation (SCBC), was formed to broadcast news and information in English. Cameroon was formed from a French colony and part of a British one in 1961 (Feldman-Savelsberg, 2001).

In 2018, Isaac approached SCBC to start a TV programme on agriculture to help Southern Cameroonians who could no longer work as a result of the war, and the thousands of refugees who sought refuge in Nigeria. The broadcasters readily agreed. With his PhD in agriculture and rural development from the University of Wisconsin-Madison, and his roots in a Cameroonian village, Isaac was well placed to find content that farmers back home would appreciate. Isaac said:

> I did some research on the internet, and I found Access Agriculture. I liked it so much that I watched every single video.

Isaac soon started a TV programme, *Amba Farmers' Voice*, which began to air every Sunday at 4 PM, Cameroon time. It is rebroadcast several times a week to give more people a chance to watch the programme. With frequent power cuts, many are not able to tune in on Sundays.

The programme is broadcast live from Isaac's studio in Virginia. He starts with a basic introduction in West African Pidgin. Isaac explains:

> If I'm going to show a video on rabbits, I start by explaining what a rabbit is. And that we can learn from farmers in Kenya how to build a rabbit house, and to care for these animals.

After playing an Access Agriculture video on the topic (in English), Isaac comments on it in Pidgin, for the older, rural viewers who may not speak English. His remarks are carefully prepared, and based on background reading and research.

The show lasts an hour or more and allows Isaac to play several videos. *Amba Farmers' Voice* has its own Facebook and YouTube pages. While his programme is on the air, Isaac checks out the Facebook page to get an idea of how many people are watching. A popular topic like caring for rabbits may have 1000 viewers just on Facebook. But most people watch the satellite broadcast. SCBC estimates that 2–3 million people watch *Amba Farmers' Voice* in Cameroon, but many others also watch it in Nigeria, Ghana, Sierra Leone and even in some Francophone countries, like Benin and Gabon.

Some farmers reciprocate, sending Isaac pictures and videos that they have shot themselves, showing off their own experiments, adapting the ideas from the videos to conditions in Cameroon. Isaac heard from one group of 'mothers in the village' who showed how they were using urine to fertilize their maize, after watching an Access Agriculture video from Uganda.

People in refugee camps watched the video *Using Sack Mounds to Grow Vegetables*, which shows how to grow vegetables in a large, soil-filled bag.

But gunny sacks were scarce in the refugee camp, so people improvised, filling plastic bags with earth and growing tomatoes in them, so they could grow some food within the confines of the camp.

Isaac mentioned that people were installing drip irrigation after seeing the video from Benin, *Drip Irrigation for Tomato*.

'That can be expensive', I said. 'People have to buy materials.'

'Not really', Isaac answered. 'Gardeners take used drink bottles from garbage dumps, fill them with water, poke holes in the cap, and leave them to drip slowly on their plants.' In other words, the farmers were creatively adapting the ideas they had learned from the video.

After seeing the video *Feeding Snails* from Benin, on raising giant African snails (for high-quality meat), one young man in the Southern Cameroons stacked used tyres to make the snail pen. It's an innovation he came up with after watching the Access Agriculture video. He put two tyres in a stack, put the snails in the bottom, and fed them banana peels and other fruit and vegetable waste. Isaac told his audience, 'We don't need to buy anything. Just open your eyes and adapt. See what you can find to use.'

Solar dryers were another innovation that people adapted from the videos. To save money, they made the dryers from bamboo instead of wood, and shared one between several families. As a further adaptation, people were drying grass in the solar dryer as livestock feed. Access Agriculture has four videos on using solar dryers to preserve high-value produce like pineapples, mangoes, leafy vegetables and chillies, but none show grass drying. Isaac explains that you sprinkle a little salt on the grass as you dry it. Then, in the dry season you put the grass in water and it turns fresh again. Now he is encouraging youth to form groups so they can dry grass to store, to sell to farmers when fodder is scarce.

I was delighted to see so many local experiments just from people who watch videos on television, with no extension support.

All of this interaction, between Isaac Zama and his compatriots, the teaching, feedback and organization, is all happening on TV and online. He hasn't been to Cameroon since he started his programme. Isaac's interaction with his audience amazes me. It's a testimony to his talent, but also to the improved connectivity in rural Africa. Isaac says:

> People think that Africans don't have cell phones, but 30% of the older farmers in villages have Android phones. Their adult children, living in cities or abroad, buy phones for their parents so they can stay in touch and so they can see each other on WhatsApp.

Isaac adds that what farmers need now is an app so they can watch agricultural videos more cheaply.

Dr Isaac Zama wants to encourage other stations to broadcast farmer learning videos:

> Those videos from Access Agriculture will revolutionize agriculture in Africa in 2 or 3 years, if our national leaders would just broadcast them on TV. The farmers would try new practices themselves, just from the information they can see on the videos.

As this experience shows, farmers don't passively copy the techniques they see on videos. The videos spark farmers' imagination to experiment with new technology, and adapt it to their own circumstances.*

▣ Related videos

Dr Zama's audience in Cameroon appreciated this video from Kenya.

Environmental Alert, DAES, KENFAP and Egerton University (2016) *How to Build a Rabbit House*. Hosted by Access Agriculture. 11 minutes.
www.accessagriculture.org/how-build-rabbit-house

The titles of other videos mentioned in the above section, *Human Urine as Fertilizer, Using Sack Mounds to Grow Vegetables, Drip Irrigation for Tomato, Feeding Snails, Solar Drying Pineapples, Making Mango Crisps, Solar Drying of Kale Leaves* and *Solar Drying of Chillies*, are available on www.accessagriculture.org.

A version of the section above was previously published on www.agroinsight.com in 2021, by Jeff Bentley.

New Crops for Mr Mpinda

A good video, one that lets farmers tell about their innovations, can spark the viewers' imagination. A video can even convince smallholders to try a new crop.

Lester Mpinda is an enterprising farmer in Mwanza, Malawi. Mr Mpinda has a vegetable garden, known as a *dimba*, which is irrigated with water from a hand-dug well. A *dimba* is hard work, but worth it.

Mr Mpinda grows vegetables and sells them in the market in Mwanza. In 2013, he was able to use his earnings to buy a small gasoline-powered pump to water his beans, onions and tomatoes. A US$100 pump is a major investment for a Malawian smallholder, but also a great way to save time and avoid the backbreaking labour of carrying water from the well to the plants during the long, hot dry season.

With the money earned from his productive garden, Mr Mpinda bought a small stand, where his wife sells vegetables in the village.

In June 2015, Ronald Kondwani Udedi left some DVDs of videos at a government telecentre managed by Mathews Kabira, near Mwanza, Malawi. The DVDs included learning videos for farmers about growing rice and chilli peppers and managing striga, the parasitic weed.

Mathews explained that he took one set of DVDs to Mr Mpinda, because he was 'a successful farmer'. Mr Mpinda had a DVD player, but no TV, so he watched the videos on chilli growing at a neighbour's house, using the neighbour's TV and Mr Mpinda's DVD player. He watched the videos as often as the neighbour would let him. The more he watched, the more he learned.

Mr Mpinda soon recognized the possibilities of chilli as a crop, even though he had never grown it.

To start a new crop, you need more than a bright idea; you need seed. Getting chilli seed took some imagination. Mr Mpinda went to the market and bought 20 small fresh chillies for 100 Kwacha (US$0.14) and then dried them, like tomatoes, and planted the little seeds in a nursery, just like he had seen on the video. Mr Mpinda was already making seedbeds for onions and some of his other vegetables. At 21 days he transplanted the chilli seedlings.

Now Mr Mpinda has several dozen chilli plants, of a perennial variety which is eaten fresh in Malawi. People cut up the fiery chilli at table, to add some zest to meals.

Every few days, Mr Mpinda harvests 3 or 4 kg of chillies and takes them to the market and sells them for 1000 Kwacha a kilogram (US$1.40).

Mr Mpinda has already planned his next step. After harvesting his little patch of aubergine (eggplant), he is going to clear the land and plant a whole garden of chilli.

Mr Mpinda has also watched the DVD of rice videos, and although no one in the area grows rice, he realizes that the crop would do well in the slightly higher space just above his rows of vegetables. He has already looked for rice seed: there is none to be found in Mwanza and the agro-dealers won't or can't order it for him, so he is going to travel to the city of Zomba, 135 km away, and buy rice seed there. Mr Mpinda has already identified the major rice varieties grown in Malawi and decided that one of them, Apasa, is the best for highland areas like his.

He became one of the first rice farmers in Mwanza district.

Mr Mpinda didn't watch the rice and chilli videos as part of a farmer group. He didn't have an extensionist to answer questions. He simply had the videos which he could (and did) watch several times to study the content. And this information alone was enough to inspire him to experiment with two crops that were entirely new to him.*

■◀ Related video

Mr Mpinda discusses his work in this video.

Udedi, K. (2016) *A New Crop for Mr Mpinda*. Hosted by EcoAgTube. 6 minutes. www.ecoagtube.org/content/new-crop-mr-mpinda

---

A version of the section above was previously published on www.agroinsight.com in 2016, by Jeff Bentley.

Malawi Calling

I was at home in Bolivia in 2018 when I got a surprise call from southern Africa. 'I'm a chilli farmer in Malawi; you've been to my house', said the confident voice on the other end, before the caller ran out of credit and the faint, crackling connection was suddenly cut off.

But the caller, Lester Mpinda, was not easily discouraged. In the time it takes to walk to the village shop and buy a scratch card, he was back on the phone. 'I've made a lot of profit from chilli', he said. Then the call was cut off again.

I remembered Mr Mpinda well. Malawian media expert Ronald Udedi and I had visited Mr Mpinda's garden 2 years previously, in September of 2016, in Mwanza, southern Malawi, where he showed us how he had started growing local chillies from seed he bought in the market after watching the videos on a DVD. I wanted to learn more, but the phone connection was too poor to chat. Instead, I contacted my friend Ronald on social media and asked him to find out more.

Ronald filled me in on the rest of Mr Mpinda's story. Shortly after our visit to his farm in 2016, Ronald and I made a short video on Mr Mpinda. Access Agriculture then invited Mr Mpinda to share his story at a meeting with partner organizations in Lilongwe, the capital city of Malawi. I couldn't attend, but I was a little apprehensive about the outcome, thinking that the event might distract Mr Mpinda from his everyday work on the farm. I couldn't have been more wrong.

At the meeting, Mr Mpinda met Mr Dyborn Chibonga, then the head of the National Smallholder Farmers' Association of Malawi (NASFAM). Mr Chibonga put Mr Mpinda in touch with the nearest NASFAM extension agent in Mulanje, who later visited the farm and gave Mr Mpinda some seed of bird's-eye chilli, the variety used to make Tabasco-style hot sauce. The slender red bottles of hot sauce are a common sight on Malawian tables and the dried chilli is exported to food-makers in Europe and elsewhere.

Chilli seed is really small, and a little bit goes a long way, so Mr Mpinda decided to share his generous gift from NASFAM with his neighbours. Mr Mpinda started a chilli club with 12 members, including eight women. He showed the club members how to plant the chilli, gave them seed, and once or twice a week he invited the club to his home to show them the chilli videos in Chichewa, the local language. Each member learned more about growing and drying this crop, which was entirely new to them. The club members created a chilli demonstration garden, where they tried out what they saw in the videos.

When the club had a stock of dried chillies, they phoned the NASFAM extension agent, who came from Mulanje, where NASFAM has a factory for making hot sauce. The agent bought 160 kg of chilli from the individual club members, paying 2500 Kwacha (US$3.50) per kilogram, twice the price of tobacco, which is the number one export crop. The NASFAM agent left more seed.

Other friends and neighbours who heard of this success asked to join the club. Mr Mpinda graciously welcomed them, and now there are 80 members growing chilli and learning about the crop from the videos.

As Ronald puts it, 'the most important thing (for starting the chilli club) was the DVD with the chilli videos. Mr Mpinda and his friends watched it to learn about everything, from taking care of the nursery beds to transplanting and harvesting.' The videos meant that farmers didn't have to rely on visits from extension agents, whose time and travel budgets are limited.

For many years, only one company, Nali, made hot sauce in Malawi, but now there are over ten manufacturers. Malawi is now enjoying a kind of chilli boom. Mr Mpinda's story shows that smallholders can independently identify and respond to market openings. Smallholder farmers are always open to new opportunities and eager to try useful innovations (Bentley *et al.*, 2010). I have no idea how long the chilli boom in Malawi will last, but agriculture will never go out of style. As long as smallholders have buyers, seed and good information, they will be able to market quality produce.*

◼️📹 Related video

Here is one of the chilli videos that inspired Mr Mpinda.

NASFAM (2016) *Drying and Storing Chillies*. Hosted by Access Agriculture. 11 minutes. www.accessagriculture.org/drying-and-storing-chillies

---

A version of the section above was previously published on www.agroinsight.com in 2018, by Jeff Bentley and Ronald Kondwani Udedi.

United Women of Morochata

The success of a woman's group depends in large part on the quality of leadership, as I saw in 2018 in Morochata, a highland municipality in the Bolivian Andes. My agronomist friend Rhimer Gonzales had organized women's groups in two neighbouring villages. One group was largely inactive, while the one in the village of Piusilla was going strong.

Rhimer phoned Juliana García, the president of the women's group of Piusilla, to arrange a meeting. Rhimer had some group business to discuss, and he was going to help me ask some follow-up questions about videos. The previous year, the women had received DVDs with seven videos on soil conservation, and I wanted to learn what the women had done with the information. Doña Juliana was not at home, and the women in her group were busy, but she said that if we came back at 8:30 that evening she would have at least some of the women at her house.

By 8 o'clock in the evening it was dark and raining hard. At 3350 m above sea level, it gets cold when it rains, and it's miserable to get wet. Rhimer and I were sure that no one would come to the meeting, but still we wanted to try.

We were surprised when we got to doña Juliana's house to see about half of the women's group there. Doña Juliana had taken the time (and spent money) to ring the women up, and had then built a warm fire to welcome them. They soon invited me to ask my questions. The videos included one that Agro-Insight made in 2017 on lupins, edible legumes that improve the soil.

The women said that they had seen the videos with Rhimer at one of their meetings. Afterwards, the women arranged to watch the videos again, by themselves, because they were looking for ways to improve their income, for example by growing lupins and broad beans. They also wanted to consolidate their position as a women's group within the *sindicato*, the local organization that represents and leads the community, but which is led mainly by men.

Besides the lupin video, the women's group had watched one from Vietnam about making live barriers on steep hillsides to conserve the soil. They recalled, accurately, that the video showed how to measure rows to plant the grass, which had to be transplanted in small clumps or cuttings.

When we asked if they had tried any of the ideas from the lupin video, doña Juliana said that she had learned how to select her seed. One of the key ideas from the lupin video is to remove the small and unhealthy grains, and only plant the best ones for a better harvest. Doña Juliana was impressed by the little hand screen she had seen in the video, to sort the grains by size, but she didn't have a screen. Instead, she just sorted the seed by hand, a practice which is also shown in the video. It is important to give people different options.

She had planted the seed, which produced a healthy crop, which was flowering during our visit. Doña Juliana was impressed that selecting her lupin seed had resulted in the plants being bigger and healthier than in previous years.

Rhimer and I asked how many of the other women in the group had selected seed too. One of them decided it was time for some comic relief. She said, 'My husband just grabbed some of the lupin grains in the bag and scattered them, and they are doing just fine.'

All of the women laughed, including doña Juliana, but then she reminded them, 'You have all seen how to select seed and you know how to do it. So, you should all try it.'

Leadership matters. In time, these women will notice the difference in yield between selected and unselected seed. It usually takes a while for a whole community to adopt an innovation. A useful step is to have one of the leaders adopt and share her experience experimenting with the ideas she has seen in the videos. Sparking innovation is like growing a crop: it requires someone to plant the seed.*

📹 **Related videos**

Here are the Bolivian video on lupins and the Vietnamese video on grass strips to conserve soil mentioned in this section. Both are also available in Quechua, among other languages.

Agro-Insight and PROINPA (2017) *Growing Lupin without Disease*. Hosted by Access Agriculture. 12 minutes.
www.accessagriculture.org/growing-lupin-without-disease

Agro-Insight (2016) *Grass Strips against Soil Erosion*. Hosted by Access Agriculture. 12 minutes.
www.accessagriculture.org/grass-strips-against-soil-erosion

---

*A version of the section above was previously published on www.agroinsight.com in 2018, by Jeff Bentley.

Call Anytime

It's difficult to know who reads a factsheet, listens to a radio broadcast or watches a farmer learning video, but those of us who produce such information always want to know what happens to it once it leaves our hands. In 2011 my colleagues at ICRISAT tried a new way to do audience research. ICRISAT and partners distributed 20,000 copies of a DVD on striga (the devil weed) across East Africa. Each copy contained a questionnaire, formatted as a letter, asking the viewers to tick off a few boxes and mail back the letter in the post. No one bothered to return the survey.

So, in 2015, then PhD candidate Gérard Zoundji tried a slightly different way to get feedback from viewers in Benin, as he explained in a 2016 paper in *Cogent Food & Agriculture* (Zoundji *et al.*, 2016). First, he compiled a DVD in five languages, with nine different videos on growing vegetables. Next, Gérard distributed his DVD through the private sector, mainly through agro-input dealers and people who sell movie DVDs. Previously DVDs had been distributed through extension providers, NGOs or government agencies, not from small shops.

Gérard asked the vendors to collect names and phone numbers of people who bought the DVD so he could do follow-up work with the buyers. Gérard gave the vendors the DVDs for free, in exchange for their cooperation, but allowed them to keep the equivalent of a dollar or two which they collected for each sale. He also tried a new way of doing follow-up. He put a sticker in the DVD jacket with a note inviting the recipients to phone in if they had questions. The number was for a SIM card that Gérard bought just to receive such calls.

It was a pleasant surprise when people started phoning in. Of 562 who bought the DVD, a whopping 341 phoned Gérard. Some just called to say how much they had enjoyed watching the videos. Others wanted to share their story. Nearly 20% of them had been so eager to watch the videos that they bought their own DVD player. Others called to ask where they could buy the drip irrigation equipment featured on one of the videos.

The six agro-input dealers who were selling the DVD were also impressed with the video on drip irrigation, and the interest it inspired in farmers. Two of these dealers actually began to stock drip irrigation supplies themselves.

As we wrote in the section 'To Drip or Not to Drip' (see Chapter 7, this volume), farmers who have been exposed to drip irrigation through development projects usually abandon the technique once the project ends. Projects usually make little effort to involve the private sector. Yet here were agro-input dealers who were motivated enough to find out where to buy the drip irrigation equipment and stock it in response to interest shown by farmers who had watched a video. Sometimes simply watching a video can excite people more than participating in a full project.

I am always intrigued to learn about someone using a cell phone in a new way, especially if it involves giving rural people the chance to make their voices heard. A sticker inside a DVD cover was enough to encourage buyers of a DVD to call in with comments.*

◼ Related video

Here is one of the vegetable videos that Gérard included in his anthology.

Agro-Insight (2016) *Insect Nets in Seedbeds*. Hosted by Access Agriculture. 12 minutes.
www.accessagriculture.org/insect-nets-seedbeds

A version of the section above was previously published on www.agroinsight.com in 2017, by Jeff Bentley.

Farmers Without Borders

Blanca Chancusig is a dynamic woman who manages a diverse farm in the village of Yugshiloma in Cotopaxi province, at 2800 m above sea level in the Ecuadorian Andes. As we film doña Blanca for a farmer-to-farmer training video, she talks about all the benefits of living hedgerows and flowering plants, which she uses not just for medicine and to feed her guinea pigs, rabbits, her two cows and a pig, but also to attract beneficial insects that pollinate and protect her crops.

Winding up our visit, Jeff and I tell Blanca that the video in which she will feature will be ready in a few months. On the ballpoint pen that we give her we point to the Access Agriculture website address, where she can find over a hundred different training videos in Spanish.

'Also on biofertilizer and good microbes', she replies. I interpret it as a question, but soon find out that it actually was a statement. She guides me to a little shed behind her house from where she brings out a plastic drum. As she opens the lid, I see she has prepared liquid biofertilizer.

'We learned a lot through the video on good microbes for the soil and the plants. It is very easy to do and not very costly. The only thing that is hard for me to get is broad bean flour, only that. Everything else I have at home', Blanca says.

Two years ago, Diego Mina and his wife, Mayra Coro, our local collaborators from the French National Research Institute for Sustainable Development (IRD – *Institut de Recherche pour le Développement*) who manage the AMIGO (Agroecological Management of crop Insects: towards a sustainable GOal for farmers) project on agroecology, showed several women's groups a video on good microbes that was made with farmers in India (and later translated into Spanish). The women liked the video so much that Diego sent them a copy of it through their WhatsApp group.

In a follow-up visit, Diego and Mayra decided to organize a live demonstration on how to make the mix of good microbes, with cow urine and raw sugar and other locally available ingredients, which the women provided. While the video shows chickpea flour as one of the ingredients, Diego and Mayra bought broad bean flour to use instead of chickpea flour, which is not available in Ecuador. The broad bean is a pulse which many of the women grow at this high altitude.

A year after the demonstration, at least some of the women are still experimenting with this new technology, which they learned through the video from fellow farmers in India. With a big smile on her face, Blanca tells me that she already tried it once on her maize and lupin crop, and that the results were good, so this planting season she has again prepared another 10 l to try on her field for a second year.

Unlike what most researchers and extensionists think, smallholder farmers deeply appreciate learning from fellow farmers, even if they are thousands of kilometres away, on other continents, especially if the technologies work well, are easy to apply and cost little money.*

📹 Related video

The Spanish version of the video on good microbes, discussed in the above section.

Pagar, A. and WOTR (2018) *Buenos Microbios para Plantas y Suelo*. Hosted by Access Agriculture. 14 minutes.
www.accessagriculture.org/es/buenos-microbios-para-plantas-y-suelo

---

A version of the section above was previously published on www.agroinsight.com in 2022, by Paul Van Mele.

References

Bentley, J.W., Van Mele, P. and Acheampong, G.K. (2010) Experimental by nature: Rice farmers in Ghana. *Human Organization* 69, 129–137.

Feldman-Savelsberg, P. (2001) Cameroon. In: Ember, M. and Ember, C.R. (eds) *Countries and Their Cultures*. Macmillan, New York, pp. 382–396.

Van Mele, P., Mohapatra, S., Tabet, L. and Flao, B. (2024) *Young Changemakers: Scaling Agroecology Using Video in Africa and India*. Access Agriculture, Brussels, Belgium.

Zoundji, G.C., Okry, F., Vodouhê, S.D. and Bentley, J.W. (2016) The distribution of farmer learning videos: Lessons from non-conventional dissemination networks in Benin. *Cogent Food & Agriculture* 2, 1277838.

Conclusion

In this book, we have explored some of the complexities facing a world that needs to change the way it feeds itself. For over a century, profit-seeking food systems have eroded local resources and cultures. Industrial agriculture has promised abundance, but the hidden costs of climate change, pollution and loss of biodiversity and of soil have come back like a boomerang, with the poorest people paying the highest price.

Democratic food systems are owned and managed by as many people as possible, not by distant corporations. Food sovereignty, climate justice and farmers' movements are all calling upon our leaders to take up their responsibilities and put measures in place to fix the broken global food system.

Agroecology can help to repair damaged soils, sequester carbon and produce healthy, local food while respecting farmers, their communities and the environment. Drawing upon our personal experiences and travels across the world, we have presented inspiring examples from Europe, Latin America, Africa and Asia. There is hope for the future. But celebrating the work of the many champions is not enough. They need to be amplified.

Smallholder farmers and others involved in building local food systems need new information if they are to successfully embrace agroecological principles and practices. In most countries, extension and advisory service providers do not have time and resources to visit all the farmers interested in learning about better ways to produce, process and sell food. Fortunately, many of these farmers now own a cell phone, which they can use to download videos and teach themselves. Extensionists, educators, entrepreneurs and media people are also showing videos, in villages, schools and universities, and on TV and via social media.

Farmers who learn new ideas about agroecology proactively use the most relevant ideas to conduct their own experiments. New ideas from videos help smallholders to creatively adapt to changes in markets and in climate.

We hope that this book has given the reader some new insights into our planet's troubled food system, and what can be done to remedy it. We are all consumers of food, fibre and beverages. By supporting the farmers who have gone the extra mile to produce healthy food without agrochemicals and with due respect for the environment and local cultures, we can make this world a better place.

Appendix: List of Videos

Videos are listed in order of appearance in the particular chapter.

Chapter 1

Agro-Insight and PROINPA (2018) *Living Windbreaks to Protect the Soil*. Hosted by Access Agriculture. 15 minutes. www.accessagriculture.org/living-windbreaks-protect-soil

Agro-Insight (2016) *Growing Cassava on Sloping Land*. Hosted by Access Agriculture. 12 minutes. www.accessagriculture.org/growing-cassava-sloping-land

AMEDD (Association Malienne d'Éveil au Développement Durable) (2016) *Making a Condiment from Soya Beans*. Hosted by Access Agriculture. 10 minutes. www.accessagriculture.org/making-condiment-soya-beans

Countrywise Communication and CIS Vrije Universiteit Amsterdam (2016) *Sustainable Land Management: Introduction*. Hosted by Access Agriculture. 7 minutes. www.accessagriculture.org/slm00-introduction

AfricaRice, Agro-Insight, IER, Intercooperation and Jekassy (2008) *Effective Weed Management in Rice*. Hosted by Access Agriculture. 15 minutes. www.accessagriculture.org/effective-weed-management-rice

Agro-Insight, IRD and INIAF (2022) *Flowering Plants Attract the Insects that Help Us*. Hosted by Access Agriculture. 15 minutes. www.accessagriculture.org/flowering-plants-attract-insects-help-us

Pagar, A. (2018) *Hydroponic Fodder*. Hosted by Access Agriculture. 14 minutes. www.access-agriculture.org/hydroponic-fodder

Siddique, R.K. (2022) *Using Good Microorganisms in Cattle Farming*. Hosted by Access Agriculture. 13 minutes. www.accessagriculture.org/using-good-micro-organisms-cattle-farming

Pagar, A. and Anthra (2021) *Natural Ways to Keep Chickens Healthy*. Hosted by Access Agriculture. 9 minutes. www.accessagriculture.org/natural-ways-keep-chickens-healthy

Chapter 2

Nawaya (2017) *Making Pressed Dates*. Hosted by Access Agriculture. 12 minutes. www.access-agriculture.org/making-pressed-dates

AMEDD (2017) *Storing Fresh and Dried Tomatoes*. Hosted by Access Agriculture. 13 minutes. www.accessagriculture.org/storing-fresh-and-dried-tomatoes

Nalunga, J. (2023) *Rearing Crickets for Food and Feed*. Hosted by Access Agriculture. 14 minutes. www.accessagriculture.org/rearing-crickets-food-and-feed

Green Adjuvants (2022) *Herbal Pest Repellent*. Hosted by Access Agriculture. 15 minutes. www.accessagriculture.org/herbal-pest-repellent

AMEDD (2016) *Helping Women Recover after Childbirth*. Hosted by Access Agriculture. 13 minutes. www.accessagriculture.org/helping-women-recover-after-childbirth

Chapter 3

Agro-Insight (2016) *Pure Milk is Good Milk*. Hosted by Access Agriculture. 8 minutes. www.accessagriculture.org/pure-milk-good-milk

Udedi, R. (2019) *Village Savings and Loan Associations*. Hosted by Access Agriculture. 14 minutes. www.accessagriculture.org/village-savings-and-loan-associations

Pagar, A. and WOTR (2020) *Organic Biofertilizer in Liquid and Solid Form*. Hosted by Access Agriculture. 15 minutes. www.accessagriculture.org/organic-biofertilizer-liquid-and-solid-form

Shanmuga Priya J. (2024) *A Hand Weeder and a Ridge Plough*. Hosted by Access Agriculture. 15 minutes. www.accessagriculture.org/hand-weeder-and-ridge-plough

Agro-Insight and PROINPA (2019) *The Wasp that Protects Our Crops*. Hosted by Access Agriculture. 9 minutes. www.accessagriculture.org/wasp-protects-our-crops

AfricaRice and Real2Reel (2016) *Rotary Weeder*. Hosted by Access Agriculture. 18 minutes. www.accessagriculture.org/rotary-weeder

Agro-Insight (2023) *Community Seed Banks*. Hosted by Access Agriculture. 15 minutes. www.accessagriculture.org/community-seed-banks

Pagar, A. (2022) *Home Delivery of Organic Produce*. Hosted by Access Agriculture. 15 minutes. www.accessagriculture.org/home-delivery-organic-produce

Chapter 4

Agro-Insight and Yanapai (2022) *Improved Pasture for Fertile Soil*. Hosted by Access Agriculture. 15 minutes. www.accessagriculture.org/improved-pasture-fertile-soil

Agro-Insight and PROSUCO (2019) *Recording the Weather*. Hosted by Access Agriculture. 14 minutes. www.accessagriculture.org/recording-weather

Agro-Insight and PROINPA (2023) *Seeing the Life in the Soil*. Hosted by Access Agriculture. 9 minutes. www.accessagriculture.org/seeing-life-soil

AMEDD (2017) *Making a Condiment from Soya Beans*. Hosted by Access Agriculture. 10 minutes. www.accessagriculture.org/making-condiment-soya-beans

DEDRAS (2018) *Growing Annual Crops in Cashew Orchards*. Hosted by Access Agriculture. 9 minutes. www.accessagriculture.org/growing-annual-crops-cashew-orchards

DEDRAS (2020) *Making Soya Cheese*. Hosted by Access Agriculture. 9 minutes. www.access-agriculture.org/making-soya-cheese

DEDRAS, Nawaya and Songhaï Centre (2016) *Feeding Grasscutters*. Hosted by Access Agriculture. 9 minutes. www.accessagriculture.org/feeding-grasscutters

AMEDD (2022) *Harvesting and Storing Okra*. Hosted by Access Agriculture. 12 minutes. www.accessagriculture.org/harvesting-and-storing-okra

Chapter 5

Agro-Insight (2017) *Farmers' Rights to Seed: Experiences from Malawi*. Hosted by Access Agriculture. 16 minutes. www.accessagriculture.org/farmers-rights-seed-experiences-malawi

Agbangla, A. (2020) *Making a Good Okra Seeding*. Hosted by Access Agriculture. 12 minutes. www.accessagriculture.org/making-good-okra-seeding

Agro-Insight (2016) *How to Make a Fertile Soil for Onions*. Hosted by Access Agriculture. 7 minutes. www.accessagriculture.org/how-make-fertile-soil-onions

Agro-Insight and Yanapai (2022) *Recovering Native Potatoes*. Hosted by Access Agriculture. 16 minutes. www.accessagriculture.org/recovering-native-potatoes

Chapter 6

Wright, C. (2018) *Living Soil*. Soil Health Institute. Hosted by EcoAgTube. 60 minutes. https://ecoagtube.org/content/living-soil

Agro-Insight (2012) *Contour Bunds*. Hosted by Access Agriculture in about 40 languages. 15 minutes. www.accessagriculture.org/contour-bunds

Agro-Insight (2015) *Grass Strips against Soil Erosion*. Hosted by Access Agriculture, in more than 20 languages. 12 minutes. www.accessagriculture.org/grass-strips-against-soil-erosion

Cota, R. (2021) *Alejandro Bonifacio Flores*. Hosted by EcoAgTube. 10 minutes. https://ecoagtube.org/content/alejandro-bonifacio-flores

Agro-Insight (2012) *Reviving Soils with Mucuna*. Hosted by Access Agriculture. 14 minutes. www.accessagriculture.org/reviving-soils-mucuna

BIID, CARE, DAM and Shushilan (2014) *The Wonder of Earthworms*. Hosted by Access Agriculture. 13 minutes. www.accessagriculture.org/wonder-earthworms

Pagar, A. and WOTR (2018) *Good Microbes for Plants and Soil*. Hosted by Access Agriculture. 14 minutes. www.accessagriculture.org/good-microbes-plants-and-soil

Priya, S. (2020) *Vermiwash: An Organic Tonic for Crops*. Hosted by Access Agriculture. 13 minutes. www.accessagriculture.org/vermiwash-organic-tonic-crops

Pagar, A. and WOTR (2020) *Better Seed for Green Gram*. Hosted by Access Agriculture. 14 minutes. www.accessagriculture.org/better-seed-green-gram

Chapter 7

Pagar, A. and WOTR (2018) *Mulch for a Better Soil and Crop*. Hosted by Access Agriculture. 12 minutes. www.accessagriculture.org/mulch-better-soil-and-crop

Green Adjuvants (2021) *Coir Pith*. Hosted by Access Agriculture. 14 minutes. www.accessagriculture.org/coir-pith

Agro-Insight (2013) *Drip Irrigation for Tomato*. Hosted by Access Agriculture. 12 minutes. www.accessagriculture.org/drip-irrigation-tomato

Pagar, A. and WOTR (2022) *Staking Tomato Plants*. Hosted by Access Agriculture. 13 minutes. www.accessagriculture.org/staking-tomato-plants

Agro-Insight and AGRECOL Andes (2023) *A Participatory Guarantee System*. Hosted by Access Agriculture. 11 minutes. www.accessagriculture.org/participatory-guarantee-system

Chapter 8

Agro-Insight (2016) *Promoting Weaver Ants in Your Orchard*. Hosted by Access Agriculture. 13 minutes. www.accessagriculture.org/promoting-weaver-ants-your-orchard

Agro-Insight (2016) *Weaver Ants against Fruit Flies*. Hosted by Access Agriculture. 11 minutes. www.accessagriculture.org/weaver-ants-against-fruit-flies

Agro-Insight and FAO (2018) *Killing Fall Armyworms Naturally*. Hosted by Access Agriculture. 16 minutes. www.accessagriculture.org/killing-fall-armyworms-naturally

CCDB (2019) *Managing Aphids in Beans and Vegetables*. Hosted by Access Agriculture. 9 minutes. www.accessagriculture.org/managing-aphids-beans-and-vegetables

Siddique, R.K. (2022) *Healthier Crops with Good Microorganisms*. Hosted by Access Agriculture. 16 minutes. www.accessagriculture.org/healthier-crops-good-micro-organisms

MSSRF (2019) *Managing Mealybugs in Vegetables*. Hosted by Access Agriculture. 11 minutes. www.accessagriculture.org/managing-mealybugs-vegetables

DEDRAS (2016) *Soya Sowing Density*. Hosted by Access Agriculture. 11 minutes. www.access-agriculture.org/soya-sowing-density

Agro-Insight, CBARDP, ICRISAT and KNARDA (2016) *Growing Row by Row*. Hosted by Access Agriculture. 9 minutes. www.accessagriculture.org/grow-row-row

Chapter 9

Pagar, A. and Anthra (2023) *Herbal Medicine against Foot Rot in Livestock*. Hosted by Access Agriculture. 12 minutes. www.accessagriculture.org/herbal-medicine-against-foot-rot-livestock

Agro-Insight (2016) *Keeping Milk Free from Antibiotics*. Hosted by Access Agriculture. 9 minutes. www.accessagriculture.org/keeping-milk-free-antibiotics

See other videos on the Access Agriculture animal health category: www.accessagriculture.org/category/133/animal-health

Priya, S. (2019) *Deworming Goats and Sheep with Herbal Medicines*. Hosted by Access Agriculture. 11 minutes. www.accessagriculture.org/deworming-goats-and-sheep-herbal-medicines

Agro-Insight and Yanapai (2022) *Improved Pasture for Fertile Soil*. Hosted by Access Agriculture. 16 minutes. www.accessagriculture.org/improved-pasture-fertile-soil

NASFAM, NOGAMU, Egerton University, ATC/UNIDO and Songhaï Centre (2016) *Taking Care of Local Chickens*. Hosted by Access Agriculture. 10 minutes. www.accessagriculture.org/taking-care-local-chickens

Egerton University (2016) *Hand Milking of Dairy Cows*. Hosted by Access Agriculture. 9 minutes. www.accessagriculture.org/hand-milking-dairy-cows

Philippine Agroforestry Education and Research Network (2024) *Raising Pigs with No Smell and Less Work*. Hosted by Access Agriculture. 13 minutes. www.accessagriculture.org/raising-pigs-no-smell-and-less-work

Chapter 10

Agro-Insight and EkoRural (2022) *Inspiring Women Leaders*. Hosted by Access Agriculture. 14 minutes. www.accessagriculture.org/inspiring-women-leaders

Agro-Insight and SWISSAID (2022) *Creating Agroecological Markets*. Hosted by Access Agriculture. 16 minutes. www.accessagriculture.org/creating-agroecological-markets

Pagar, A. (2022) *Home Delivery of Organic Produce*. Hosted by Access Agriculture. 15 minutes. www.accessagriculture.org/home-delivery-organic-produce

Practical Action Nepal (2019) *Managing Tomato Late Blight*. Hosted by Access Agriculture. 9 minutes. www.accessagriculture.org/managing-tomato-late-blight

Agro-Insight (2016) *Taking Milk to the Collection Centre*. Hosted by Access Agriculture. 13 minutes. www.accessagriculture.org/taking-milk-collection-centre

Agro-Insight (2023) *Collecting Traditional Varieties*. Hosted by Access Agriculture. 15 minutes. www.accessagriculture.org/collecting-traditional-varieties

KENAFF, Farm Radio Trust Malawi, UNIDO/Egypt and Farmers Media (2016) *Making Banana Flour*. Hosted by Access Agriculture. 12 minutes. www.accessagriculture.org/making-banana-flour

Chapter 11

Nawaya (2017) *Making Fresh Cheese*. Hosted by Access Agriculture. 13 minutes. www.access-agriculture.org/making-fresh-cheese

DEDRAS (2018) *Growing Annual Crops in Cashew Orchards*. Hosted by Access Agriculture. 9 minutes. www.accessagriculture.org/growing-annual-crops-cashew-orchards

Agro-Insight and SWISSAID (2022) *Creating Agroecological Markets*. Hosted by Access Agriculture. 16 minutes. www.accessagriculture.org/creating-agroecological-markets

AMEDD (2017) *Tomato Concentrate and Juice*. Hosted by Access Agriculture. 12 minutes. www.accessagriculture.org/tomato-concentrate-and-juice

Tous Terriens (2021) *Paysan-Boulanger/The Farmer-Baker*. Hosted by EcoAgTube. 10 minutes. www.ecoagtube.org/content/paysan-boulanger-farmer-baker

DEDRAS (2016) *Preparing Cashew Apple Juice*. Hosted by Access Agriculture. 9 minutes. www.accessagriculture.org/preparing-cashew-apple-juice

FIBL (2021) *Marketing Organic Products in Africa*. Hosted by EcoAgTube. 58 minutes. www.ecoagtube.org/content/marketing-organic-products-africa

Agro-Insight and AGRECOL Andes (2023) *A Participatory Guarantee System*. Hosted by Access Agriculture. 11 minutes. www.accessagriculture.org/participatory-guarantee-system

Agro-Insight (2016) *Managing Vegetable Nematodes*. Hosted by Access Agriculture. 16 minutes. www.accessagriculture.org/managing-vegetable-nematodes

Chapter 12

Nawaya (2016) *Making Pressed Dates*. Hosted by Access Agriculture. 12 minutes. www.access-agriculture.org/making-pressed-dates

Agro-Insight, CIZA and IDMA (2022) *Teaching Agroecology in Schools*. Hosted by Access Agriculture. 14 minutes. www.accessagriculture.org/teaching-agroecology-schools

AMEDD (2016) *Enriching Porridge*. Hosted by Access Agriculture. 13 minutes. www.access agriculture.org/enriching-porridge

Hochschule Rhein-Waal and Biovision (2020) *Enriching Porridge with Baobab Juice*. Hosted by Access Agriculture. 8 minutes. www.accessagriculture.org/enriching-porridge-baobab-juice

Agro-Insight and AGRECOL Andes (2023) *How to Sell Ecological Food*. Hosted by Access Agriculture. 15 minutes. www.accessagriculture.org/how-sell-ecological-food

Chapter 13

Agro-Insight (2016) *Making Chilli Powder*. Hosted by Access Agriculture. 11 minutes. www.accessagriculture.org/making-chilli-powder

Environmental Alert and Farmers Media (2020) *Housing for Pigs*. Hosted by Access Agriculture. 13 minutes. www.accessagriculture.org/housing-pigs

KENAFF, FRT Malawi, AIS Egypt and Malawi Polytechnic (2020) *Making Sausages from Rabbit Meat*. Hosted by Access Agriculture. 12 minutes. www.accessagriculture.org/making-sausages-rabbit-meat

Agbangla, A. (2018) *Making Groundnut Oil and Snacks*. Hosted by Access Agriculture. 12 minutes. www.accessagriculture.org/making-groundnut-oil-and-snacks

Agro-Insight and CIAT (2019) *Managing Birds in Climbing Beans*. Hosted by Access Agriculture. 10 minutes. www.accessagriculture.org/managing-birds-climbing-beans

Chapter 14

Agro-Insight and FAO (2018) *Scouting for Fall Armyworms*. Hosted by Access Agriculture. 14 minutes. www.accessagriculture.org/scouting-fall-armyworms

Agro-Insight and AGRECOL Andes (2023) *A Participatory Guarantee System*. Hosted by Access Agriculture. 11 minutes. www.accessagriculture.org/participatory-guarantee-system

Philippine Permaculture Association, Alangilan National High School and NISARD (2024) *Record-Keeping for Integrated Farming*. 15 minutes. www.accessagriculture.org/record-keeping-integrated-farming

Shushilan, BIID and mPower (2020) *Hardening Crabs in Floating Cages*. Hosted by Access Agriculture. 10 minutes. www.accessagriculture.org/hardening-crabs-floating-cages

Biovision (2019) *Making Mango Crisps*. Hosted by Access Agriculture. 11 minutes. www.access-agriculture.org/making-mango-crisps

MSSRF (2022) *Organic Growth Promoter for Crops*. Hosted by Access Agriculture. 16 minutes. www.accessagriculture.org/organic-growth-promoter-crops

Chapter 15

Agro-Insight and ICRISAT (2016) *Striga Biology*. Hosted by Access Agriculture. 9 minutes. www.accessagriculture.org/striga-biology

Pagar, A. and WOTR (2023) *Controlling Wilt Disease in Pigeon Pea*. Hosted by Access Agriculture. 15 minutes. www.accessagriculture.org/controlling-wilt-disease-pigeon-pea

Agro-Insight and PROINPA (2023) *Managing the Potato Tuber Moth*. Hosted by Access Agriculture. 15 minutes. www.accessagriculture.org/managing-potato-tuber-moth

Agro-Insight, UMSA and PROSUCO (2019) *Forecasting the Weather*. Hosted by Access Agriculture. 15 minutes. www.accessagriculture.org/forecasting-weather

Agro-Insight (2016) *Women in Extension*. Hosted by Access Agriculture. 16 minutes. www.accessagriculture.org/women-extension

Agro-Insight and PROSUCO (2023) *Making Enriched Biofertilizer*. Hosted by Access Agriculture. 16 minutes. www.accessagriculture.org/making-enriched-biofertilizer

Chapter 16

Agro-Insight, CIP and INIAF (2022) *Managing Seed Potato*. Hosted by Access Agriculture. 16 minutes. www.accessagriculture.org/managing-seed-potato

Agro-Insight, UMSA and PROSUCO (2019) *Forecasting the Weather*. Hosted by Access Agriculture. 15 minutes. www.accessagriculture.org/forecasting-weather

NASFAM (2017) *Intercropping Maize with Pigeon Peas*. Hosted by Access Agriculture. 10 minutes. www.accessagriculture.org/intercropping-maize-pigeon-peas

Agro-Insight (2018) *Managing Aflatoxins in Maize before and during Harvest*. Hosted by Access Agriculture. 14 minutes. www.accessagriculture.org/managing-aflatoxins-maize-and-during-harvest

Agro-Insight and Yanapai (2022) *Improved Pasture for Fertile Soil*. Hosted by Access Agriculture. 16 minutes. www.accessagriculture.org/improved-pasture-fertile-soil

Agro-Insight and PROSUCO (2019) *Recording the Weather*. Hosted by Access Agriculture. 14 minutes. www.accessagriculture.org/recording-weather

Chapter 17

Agro-Insight and PROINPA (2023) *Seeing the Life in the Soil*. Hosted by Access Agriculture. 9 minutes. www.accessagriculture.org/seeing-life-soil

Agro-Insight, IRD and INIAF (2022) *Flowering Plants Attract the Insects that Help Us*. Hosted by Access Agriculture. 15 minutes. www.accessagriculture.org/flowering-plants-attract-insects-help-us

Agro-Insight and Mountain Institute (2022) *Rotational Grazing*. Hosted by Access Agriculture. 15 minutes. www.accessagriculture.org/rotational-grazing

Agro-Insight and *Fundación Valles* (2022) *Managing Aflatoxins in Groundnuts during Drying and Storage*. Hosted by Access Agriculture. 16 minutes. www.accessagriculture.org/managing-aflatoxins-groundnuts-during-drying-and-storage

AIS, MSSRF and WOTR (2016) *Preparing Low-Cost Concentrate Feed*. Hosted by Access Agriculture. 10 minutes. www.accessagriculture.org/preparing-low-cost-concentrate-feed

Agro-Insight (2016) *Succeed with Seeds*. Hosted by Access Agriculture. 11 minutes. www.accessagriculture.org/succeed-seeds

Chapter 18

Agro-Insight, ICRISAT and UACT (2016) *Integrated Approach against Striga*. Hosted by Access Agriculture. 8 minutes. www.accessagriculture.org/integrated-approach-against-striga

Agro-Insight, CABI, Countrywise Communication, IRRI, RDA and TMSS (2003) *Rice Seed Preservation*. Hosted by Access Agriculture. 7 minutes. www.accessagriculture.org/rice-seed-preservation

Agro-Insight and PROINPA (2018) *Living Windbreaks to Protect the Soil*. Hosted by Access Agriculture. 15 minutes. www.accessagriculture.org/living-windbreaks-protect-soil

Agro-Insight and PROSUCO (2023) *Making Enriched Biofertilizer*. Hosted by Access Agriculture. 16 minutes. www.accessagriculture.org/making-enriched-biofertilizer

Agro-Insight and PROINPA (2017) *Growing Lupin without Disease*. Hosted by Access Agriculture. 12 minutes. www.accessagriculture.org/growing-lupin-without-disease

AIS, MSSRF and WOTR (2016) *Stocking Fingerlings in a Nursery Pond*. Hosted by Access Agriculture. 12 minutes. www.accessagriculture.org/stocking-fingerlings-nursery-pond

Agro-Insight (2016) *Reviving Soils with Mucuna*. Hosted by Access Agriculture. 14 minutes. www.accessagriculture.org/reviving-soils-mucuna

Chapter 19

Videos on striga can be seen at: www.accessagriculture.org/search/striga/all

Malawi Broadcasting Corporation (2017) *Interview with Kondwani Udedi*. Hosted by EcoAgTube. 6 minutes. www.ecoagtube.org/content/mbc-interview

NASFAM, NOGAMU, Egerton University and ATC/UNIDO (2016) *Turning Honey into Money*. Hosted by Access Agriculture. 11 minutes. www.accessagriculture.org/turning-honey-money

Philippine Permaculture Association, Alangilan National High School and NISARD (2024) *Record-Keeping for Integrated Farming*. 15 minutes. www.accessagriculture.org/record-keeping-integrated-farming

Over 50 videos in Twi (at the time of publication): www.accessagriculture.org/search/all/tw

WOTR (2018) *Making a Vermicompost Bed*. Hosted by Access Agriculture. 16 minutes. www.accessagriculture.org/making-vermicompost-bed

Agro-Insight, AMEDD, Countrywise Communication, FLASH, Fuma Gaskiya, ICRISAT, INRAN and UACT (2016) *Composting to Beat Striga*. Hosted by Access Agriculture. 10 minutes. www.accessagriculture.org/composting-beat-striga

Agro-Insight and PROINPA (2017) *Producir Tarwi sin Enfermedad*. Hosted by Access Agriculture. 12 minutes. www.accessagriculture.org/es/producir-tarwi-sin-enfermedad

Practical Action Nepal (2018) *Making a Modern Beehive*. Hosted by Access Agriculture. 15 minutes. www.accessagriculture.org/making-modern-beehive

Chapter 20

Access Agriculture (2022) *Entrepreneurs for Rural Access (ERA): Elphas Masanga*. Hosted by EcoAgTube. 3 minutes. www.ecoagtube.org/content/entrepreneurs-rural-access-era-elphas-masanga

Environmental Alert, DAES, KENFAP and Egerton University (2016) *How to Build a Rabbit House*. Hosted by Access Agriculture. 11 minutes. www.accessagriculture.org/how-build-rabbit-house

Udedi, K. (2016) *A New Crop for Mr Mpinda*. Hosted by EcoAgTube. 6 minutes. www.ecoagtube.org/content/new-crop-mr-mpinda

NASFAM (2016) *Drying and Storing Chillies*. Hosted by Access Agriculture. 11 minutes. www.accessagriculture.org/drying-and-storing-chillies

Agro-Insight and PROINPA (2017) *Growing Lupin without Disease*. Hosted by Access Agriculture. 12 minutes. www.accessagriculture.org/growing-lupin-without-disease

Agro-Insight (2016) *Grass Strips against Soil Erosion*. Hosted by Access Agriculture. 12 minutes. www.accessagriculture.org/grass-strips-against-soil-erosion

Agro-Insight (2016) *Insect Nets in Seedbeds*. Hosted by Access Agriculture. 12 minutes. www.accessagriculture.org/insect-nets-seedbeds

Pagar, A. and WOTR (2018) *Buenos Microbios para Plantas y Suelo*. Hosted by Access Agriculture. 14 minutes. www.accessagriculture.org/es/buenos-microbios-para-plantas-y-suelo

Index

Page numbers in **bold** refer to figures and tables.

CABI – who we are and what we do

This book is published by **CABI**, an international not-for-profit organisation that improves people's lives worldwide by providing information and applying scientific expertise to solve problems in agriculture and the environment.

CABI is also a global publisher producing key scientific publications, including world renowned databases, as well as compendia, books, ebooks and full text electronic resources. We publish content in a wide range of subject areas including: agriculture and crop science / animal and veterinary sciences / ecology and conservation / environmental science / horticulture and plant sciences / human health, food science and nutrition / international development / leisure and tourism.

The profits from CABI's publishing activities enable us to work with farming communities around the world, supporting them as they battle with poor soil, invasive species and pests and diseases, to improve their livelihoods and help provide food for an ever growing population.

CABI is an international intergovernmental organisation, and we gratefully acknowledge the core financial support from our member countries (and lead agencies) including:

UKaid
from the British people

Ministry of Agriculture
People's Republic of China

Australian Government
Australian Centre for
International Agricultural Research

Agriculture and
Agri-Food Canada

Ministry of Foreign Affairs of the
Netherlands

Schweizerische Eidgenossenschaft
Confédération suisse
Confederazione Svizzera
Confederaziun svizra

Swiss Agency for Development
and Cooperation SDC

Discover more

To read more about CABI's work, please visit: **www.cabi.org**

Browse our books at: **www.cabi.org/bookshop**,
or explore our online products at: **www.cabi.org/publishing-products**

Interested in writing for CABI? Find our author guidelines here:
www.cabi.org/publishing-products/information-for-authors/

www.ingramcontent.com/pod-product-compliance
Lightning Source LLC
Chambersburg PA
CBHW050104220326
41598CB00043B/7383